Beyenan NGARASNDI

Cotton production and communication in Chad from 1928 to 2017

Beyenan NGARASNDI

Cotton production and communication in Chad from 1928 to 2017

The socio-economic and political challenges of cotton growing in Chad

ScienciaScripts

Imprint

Cover image: www.ingimage.com

This book is a translation from the original published under ISBN 978-620-6-70122-4.

Publisher:
Sciencia Scripts
is a trademark of
Dodo Books Indian Ocean Ltd. and OmniScriptum S.R.L publishing group

120 High Road, East Finchley, London, N2 9ED, United Kingdom
Str. Armeneasca 28/1, office 1, Chisinau MD-2012, Republic of Moldova, Europe
Printed at: see last page
ISBN: 978-620-8-02998-2

Contents

DEDICATION

My parents Ngarasndi Naidombal, Romneloum Madeleine and my uncle Clamra Célestin.

SUMMARY

The cotton industry was introduced in Chad during the colonial period and has been a source of considerable economic and social development. Following the introduction of cotton growing, a number of ginning factories were set up in the southern part of the country, managed by COTONFRAN in accordance with the commercial agreement established by cotton companies in Africa. As a result, cotton growing has been imposed by the authorities on the local population, with the support of traditional chiefs for its expansion. The local people had difficulty adapting to the systems and methods of cultivation, forcing the colonial administration to follow them during the agricultural and commercial campaigns. Over the years, this crop has changed the rural world to a large extent through animal traction, motorisation and the infrastructure enabling its expansion. In the structural context, the government has set up technical research training centres to support cotton growers in the Sudanian zone. These factors encouraged the development of cotton production, making Chad one of the world's leading cotton-producing countries in the 1980s. But falling world prices have changed the trend in Chad's cotton production in recent years, as compared with other countries in West and Central Africa. This is due to poor organisation by COTONTCHAD in the distribution of inputs, poor supervision, delays in the collection of seed cotton and payment to producers, as well as rising prices for inputs and agricultural equipment.
Key words: production, marketing, cotton, Mandoul Oriental, Chad.

ABSTRACT

The cotton sector entered Chad during the colonial period and is a source of considerable economic and social development. After its introduction, many factories were created in the southern part of the country managed by the company COTONFRAN according to the commercial agreement fixed by the cotton companies in Africa. Thus, cotton cultivation was imposed by the administration on local populations, supported by the traditional chiefs for its expansion. They find it difficult to adapt to the agricultural systems and methods that oblige the colonial administration to follow them during the agricultural and commercial campaigns. Over the years, this culture has changed the rural world in large part through animal traction, motorization as well as infrastructure allowing its extension. In the structural context, the administration has set up technical research training centres to supervise cotton farmers in the Sudanian zone. These factors contributed to the development of cotton production, calling Chad a major cotton producing country in the 1980s, but the decline in world prices changed the trend of Chadian cotton production down in recent years compared with other countries in West and Central Africa. This is due by poor organization of COTONTCHAD in the distribution of inputs, managerial failure, delays in seed cotton removal and payment to producers, as well as higher prices for inputs and agricultural equipment.

Key words: production, marketing, cotton, Mandoul Oriental, Chad.

GENERAL INTRODUCTION

I - Presentation of the subject

The cotton sectors play an important role in the development of certain African countries. They make a significant contribution to the gross domestic product of these countries and employ millions of people who derive most of their cash income from the sector[1] . In some countries, notably those in the Franc zone, cotton accounts for around 10% of gross domestic product[2] . Cotton growing was introduced in 1928 and made compulsory by the colonial administration. Apart from oil, cotton is Chad's leading export. It plays a vital role in the economy and provides a livelihood for more than three million Chadians, almost a third of the country's total population[3] . Agriculture is an essential part of the economy of all African countries. It plays an essential role in meeting human needs such as eradicating poverty and hunger, boosting inter-African trade and investment, rapid industrialisation and economic diversification, and sustainable management of resources and the environment[4] . This study summarises the historical and economic panorama of Africa in general, and the socio-economic development of Chad in particular. Chad remains an agricultural country, faced with food insecurity and the poor performance of its agriculture, which continues to employ around 80% of its working population and contributes almost 40% of GDP[5] . Consequently, the subject we are focusing on in this study is "**The production and marketing of cotton in the Mandoul Oriental region (Chad) from 1928 to 2016**".

II - Reasons for choosing this subject

There are two reasons for choosing this subject: scientific and personal.

On a scientific level, this work shows the dynamics of cotton production in Chad, which contributes to economic development. Cotton production in Chad has enabled the development of food crops thanks to access to fertilisers. National agricultural production is subject to a variety of climatic hazards and can vary significantly from one year to the next, creating a situation of near-permanent food insecurity. Every year, the country strives to absorb cereal products through commercial imports and food aid[6] . This is why a number of organisations are working in Chad to combat food insecurity, including the United Nations Development Programme (UNDP)[7] . It provides all the technical, material and financial assistance needed to implement the programme's action plan, with a view to helping poor countries, consolidating peace and stabilising communities in order to set themselves on the road to economic recovery and achieving the Millennium Development Goals (MDGs). This assistance was provided in accordance with the principles of the Paris Declaration and the Accra Agenda for Action on aid effectiveness, with an emphasis on building national capacities for effective participation and ownership[8] .

On a personal level, the choice of this subject is not a matter of chance, but the industrial and commercial sector has been a lever for Chad's economic development for decades, right up to

---

[1] Privatisation and liberalisation of the African cotton sectors.

[2] *Ibid.*

[3] N:/OZ/Expansion/2005/contracts/Global/GFCprojects/GFC1/Tchad/125339-Tchad-Privatisationcoton/Documents/Rapportfinal/Rap_final_cowi_3.Doc, consulted on 05 May 2017

[4] NEPAD, 2013, *Les agricultures africaines, transformations et perspectives*, November, p. 72.

[5] FAO, 2004, *Food and Agriculture Organization of the United Nations*.

[6] FAO, 2004, Programme Nationale d'Investissement à Moyen Terme (PNIMT) au Tchad, p.39.

[7] *Ibid.*

[8] *Ibid*, p. 39.

the present day. We have noted that cotton production and marketing contribute to the country's socio-economic development. In fact, this sector of activity requires a sufficient workforce and the majority of the population of the region and many others are employed in the cotton ginning plant.

III - Interest in the subject

This work deals with the production and marketing of cotton in the agro-industrial sector in southern Chad for economic and social development. The issue of agricultural production has its origins in world history, especially in Africa, which is considered to be the cradle of humanity. However, four interests underlie this work: scientific, political, social and economic.

From a scientific point of view, cotton production requires work based on experience in studying seeds suited to the soil, nitrogen, phosphorus-potassium (NPK) and urea fertilisers, pesticides, equipment and batteries (preventive measures). It should also be noted that cotton growers are supervised by the Agents Cotonniers de Terrain (ACT) set up by the Office National du Développement Rural (ONDR), the Institut de Recherche du Coton et des Textiles (IRCT) in Chad and in many countries in the Franc zone[9] .

Politically, the State has a monopoly on controlling the management of the company for the benefit of society, by participating in decision-making on the price of cotton and the organisation and development of COTONTCHAD. Cotton production enables the Chadian state to expand its economy and forge diplomatic relations with the outside world.

Socially, cotton production and marketing have played a major role in the lives of the people of Chad. This is because it enables the Chadian people to integrate socially into the market at farm level (self-managed market) and also at the ginning plant. In addition, these activities lead to social cohesion, demographic growth and the emergence of other activities in the locality, such as the creation of schools, health districts, village associations, water wells and vocational training centres.

In economic terms, cotton contributes 40% of Chad's agricultural production, despite the fall in prices, accounting for 63% of official exports in 1987 compared with 87% in 1984[10] . From then until now, cotton production and marketing have played a major role in the country's economic life, even though it has experienced crises.[11]

IV - Conceptual framework

The study of the production and marketing of the cotton sector in Chad, introduced in 1928, has played an important role in the country's socio-economic development. The following concepts should therefore be defined: production, marketing, cotton and development.

The Dictionnaire Robert alphabétique et Analogique de la langue française[12] defines production as the action of causing or producing (a phenomenon). It is the act of producing more or less (speaking of land, a business) of the goods created by agriculture or industry. Production is also the act of ensuring the conditions for the creation of economic wealth (goods, services); the stage of the economy at which production takes place. According to

[9] Groupe de travail et de Coopération Française, p. 42.

[10] R. Georges, 1989, *Le coton en Afrique de l'Ouest et du Centre*, situation et perspectives, Montpellier: CIRAD-MESRU, p. 53.

[11] P. Artus, 1998, "Les entreprises françaises vont-elles recommencer à s'endetter?", *Revue d'économie Financière*, no. 46, Endettement/Surendettement, https: //www. Org/stable/42903599 ? seq = page_scan_tab_contents, accessed on 27 March 2017.

[12] G. Dumond, 1982, *Le Robert alphabétique et analogique française*, Paris, Gallimard, p.1398.

Marxists, production is the combination of productive forces and the social relations of production. The latter refer to social relations, i.e. the relations that people have with each other in a given mode of production.
According to the Larousse dictionary[13] , production is the "transformation of a raw material into a product" by means of work. The concept of production varies from one doctrine to another: the physiocrats, the Marxists, the classics and the neo-classics. Three concepts are involved: the Factors of Production; the goods or services involved in the production of a good; and the mode of production. Production is a theoretical concept which considers the social totality as a structure in which the economic level is ultimately decisive (historical materialism). Relationships of production; apart from technical relationships, these correspond to the social relationships that develop on the basis of ownership of the means of production. They determine the division of society into classes[14] .
According to Joël Privost[15] in Les mots de l'économie, production refers to any market or non-market activity deemed to produce value. It encompasses the manufacture of goods and the provision of services. Production involves a certain number of factors, known as "production" factors. These are: labour, capital and the resources of the soil and subsoil (land). Production is in fact a combination of factors that is carried out according to the cost and expected efficiency of each of them. The author defines market production and non-market production.
Market production: goods and services actually sold on a market at a certain price. Non-market production, on the other hand, refers to goods and services that satisfy the needs of consumers outside the market (i.e. not sold) or that are sold at a price lower than their real cost. This is the case, for example, of most services provided by administrators.
The Dixeco of economics[16] defines production as the quantity of goods produced during a manufacturing cycle or in a given time. On the market, production becomes supply. It results from the combination of several factors: raw materials, labour and capital, known as the factors of production.
Commercialisation, according to the Larousse dictionary,[17] is the action of marketing. To market is to put on the market, to launch, to develop the commercial distribution of a product. In the same logic as the word marketing, the word commercial appears, which means to make the object of a trade.
Commerce: from the Latin "commercium", is an operation whose purpose is the sale of a commodity, a value, or the purchase of it in order to return it after having transformed it or not; the company that carries out this operation. This involves trading, buying, selling, circulating, transiting, transporting, banking and foreign exchange[18] . It is an exchange activity that can be understood at two levels: the nation and the world, i.e. national and international trade. At the national level, the study of trade relates directly to the study of sales and distribution networks that link individuals and producers through the exchange of goods. At the international level, trade reveals both relations between nations and the globalisation of

[13] J. Pierre, 2008, *Larousse illustré*, 6ème edition, L'Harmattan, p. 824.
[14] M. Grawitz, 2004, *Lexique des sciences sociales*, 8th edition, Dalloz, p.327.
[15] J. Privost, 1986, *Les mots de l'économie*, Paris, L'Harmattan, p. 301.
[16] C. Coch, 1984, *Dixeco de l'économie*, 1ère edition, Paris, Gallimard, p. 83.
[17] M. Grandmangin, 1987, *Dictionnaire Larousse*, Paris, Gallimard, p. 225.
[18] A. Maurois, 1937, Dictionnaire de l'Académie français, 8ème edition, Mercure de France, https://fr.wikionary.org/wiki/commerce, accessed 30 March 2017.

the economy and the increasing loss of control by states[19] .

Cotton: derived from the Arabic *qutuun,* is a plant fibre that surrounds the seeds of "true" cotton plants (*Gossypium Sp.*), a shrub in the Malvaceae family. This fibre is generally transformed into yarn, which is woven into fabrics[20] . Cotton is the fruit of the cotton plant, an herbaceous plant or shrub native to India, with yellow or pink flowers, cultivated in warm regions to produce cotton, grains and oil. There are four species of cotton plant: *Gossypium hirsutum*, which has hairy leaves and white to cream flowers from November to May. Its fruits are round, smooth capsules with stamens and a fairly short stigma. *Gossypium berbaceum* is the cotton variety with pale yellow flowers in July and August and small, round fruit capsules. *Gossypium arboreum*, which has leaves with very pronounced rounded lobes, pale yellow flowers and elongated capsule fruits. *Gossypium barbadense* has smooth leaves and flowers with long stigmas protruding from the stamens. There are red spots at the base of the yellow petals and the fruits are elongated capsules[21] . Cotton is a cash crop in Africa in general and in Chad in particular during the colonial period, but was not known to the people south of the Lake Chad basin until the 19th century ème[22] .

Development: according to the Grand Robert dictionary, development is the act of growing. It refers to growth and fulfilment.[23] Serge Latouche sees the concept of development as an invention of the West to standardise the world in its image. He finds that for the Third World, the term refers to economic colonisation, inequality and the destruction of the environment and culture. Latouche gives the word development various adjectives: self-centred development, endogenous development, participatory development, community development, equitable development, micro-development and the 24
local development.[24]

François Perroux defines development as "the combination of mental and social changes that enable a nation to achieve cumulative and lasting growth in its overall real product"[25] .

Rural development: this refers to the management of human development and the orientation of technological and institutional change to improve inclusion, longevity, knowledge and living standards in rural areas, in a context of equity and sustainability[26] .

Sustainable development: this consists of harmonious exploitation of the environment while preserving its continued degradation in human society[27] . Jean Marc Ela believes that, for governments, sustainable development seems to involve the cultivation of cotton or cash

[19] The Editions Larousse website, www.larousse.fr./encyclopedie/divers/commerce/35477, accessed on 13 March 2017.

[20] File://D:/coton-wikipedia.htm, accessed on 13 March 2017.

[21] Anon, 1991, *Mémento de l'agronome*, Ministère de la Coopération et du Développement, 4e Edition, Paris, Collection " Techniques Rurales en Afrique ", quoted by Djoubdje Alamine, 2013, " L'impact de la culture du coton dans la Tandjile-Est (Tchad) :1930-2012 ", Mémoire de Master en Histoire, Université de Ngaoundéré, p.85.

[22] G. Magrin, 2001, "Le sud du Tchad en mutation : des champs du coton aux sirènes de l'or noir", PhD thesis in geography, Université de Paris, Panthéon-Sorbonne, p.82-84.

[23] Josette Rey-Debove, 1999, *Le dictionnaire grand robert*, CLE international 27, rue de la Glacière, Paris XIIIe , p 129.

[24] S. Latouche, 1989, *L'occidentalisation du monde : Essai sur la signification, la portée et les limites de l'uniformisation planétaire*, Paris, la découverte. en.m.wikipedia.org/wiki/File: serge-latouche.jpg, consulted 13 March 2017.

[25] F. Perroux, 1964, *L'économie du XXème siècle*, Paris, PUF, p. 62.

[26] https://www.aquaportail.com/definition_5841_developpement rural.htm, consulted on 30 June 2017

[27] https://fr.wikipedia.org/wiki. Histoire de_la_culture du coton, accessed on 30 June 2017

crops, which is why the state deploys its agents to train cotton growers[28] .
Marketing: all the studies and actions involved in creating products that satisfy the needs and desires of consumers and ensure that they are marketed in the most profitable conditions. It includes all the activities that take products from the producer to the consumer. In addition to sales, these activities include functions such as purchasing, transport, warehousing, finance and advertising[29] .
According to B. Bathelot, marketing is the sum total of actions aimed at studying and influencing consumer needs and behaviour. It makes it possible to continuously adapt production and the commercial apparatus in line with previously identified needs and behaviour[30] .
Transport: from the Latin *trans*, meaning beyond, and *partare*, meaning to carry. Transport is the act of carrying something or someone from one place to another. It includes the notion of vehicle and communications route (road, canal). Communication routes are part of the transport infrastructure, as are the engineering structures (bridges, tunnels) and buildings (stations, car parks) associated with them[31] .

V - Theoretical framework

A theory is defined as a set of laws drawn up by thinkers to apply to a given field or phenomenon. As far as our work is concerned, many theories have been developed relating to the production and marketing of cash crops in the world in general and in Africa in particular. To carry out this study, therefore, we have drawn on theories relating to the historical, economic, sociological and political approaches.
The classical theory of international trade developed by Adam Smith gave rise to a debate on the comparative advantage model developed by David Ricardo in 1817. In the 18th century, international trade was opposed to the mercantilist doctrine. It was the subject of scientific analyses initiated by the founder of modern economics, Adam Smith. His work aimed to show that trade between nations brings a net gain to the trading nations. According to him, a product is only exported when it has lower costs and higher productivity than competing products.
The British economist Adam Smith, the founder of modern economics, was a resolute champion of free markets and free trade, and his arguments are indisputable: free trade enables countries to benefit from their comparative advantages; all countries win when each specialises in the areas in which it excels[32] .
The international trade model developed by Adam Smith is based on the notion of absolute advantage, the principle of specialisation and the international division of labour. The notion of absolute advantage assumes that countries have abundant initial endowments of natural resources or technological progress in certain sectors of activity. The principle of specialisation stipulates that each country must specialise in those areas of activity in which it

[28] J-M. Ela, 1994, Afrique : l'irruption des pauvres. Société contre l'Ingérence, pouvoir et Argent, L'Harmattan, Paris, p.100. Quoted by Djoubdje Alamine, 2013, "L'Impact de la culture du coton dans la Tandjile-Ouest (Tchad): 1930-2012", Université de Ngaoundéré, p.9.
[29] J. C. Chebat and B. G. Simard, 1971, *Le vendeur méconnu dans connu*, www.cnrt/en/definition/marketing, consulted on 26 June 2017
[30] B. Bathelot, 2015, "L'encyclopédie illustrée du marketing", https://www.google.com/search?q=marketing&ie=utf-8a&oe=utf-8, consulted on 26 June 2017
[31] Transport website - Ministry of Ecology, Sustainable Development and Town and Country Planning: http://www.transports.equipement.gouv.fr, consulted on 26 June 2017
[32] E. J. Stiglitz and A. Charlton, 2005, *"Pour un commerce plus juste"*, Paris, PUF, p.497.

does not have an absolute advantage.
According to Adam Smith, the international division of labour benefits all trading countries in the world and leads to an optimal allocation of labour. While it is true that, from a theoretical and analytical point of view, Adam Smith's model of the international specialisation of labour makes it possible to explain certain advantages of opening up to autarky, it has to be acknowledged that, empirically, this approach remains unsatisfactory. Many nations continue to export products for which they have no absolute advantage. In addition, there are countries that are net importers of all products. These criticisms led David Ricardo to revisit the concept of absolute advantages with an approach based on relative advantages. Ricardo challenged Adam Smith's theory of absolute advantages by explaining that reasoning should be based not on absolute costs but on relative costs. According to Boussard[33] , David Ricardo is the true author of the justification for trade between nations that economists have never provided. Based on the theory of relative or comparative advantage in terms of the endowment of natural factors or technological advances, David Ricardo showed that international trade is mutually advantageous if each country specialises in production where it has an absolute cost advantage. Although close to Adam Smith's idea, the Ricardian model marks a major difference.
Ricardo showed in 1817 that all countries, even those whose unit production costs are higher than in other countries for all goods, nevertheless have an interest in specialising in the production of goods for which their relative disadvantage is the least[34] . However, the concept of relative advantage stipulates that a country has a relative comparative advantage over another country in production where its cost is the least distant from that of the most comparative country, i.e. in production where the difference in cost between the two is the smallest. According to Smith and Ricardo, international trade rests on the principle of the international division of labour based on the comparative advantages of nations with a view to satisfying needs using fewer factors of production while benefiting all the co-trading countries. The theory of international trade is very important in the context of this study insofar as the cotton production of African countries has been the subject of exports of cotton fibre on a global scale. These countries have benefited from the cost of their products, as have the importing countries.
Duopoly theory of competition shows that a firm's industrial and commercial choices[35] have an impact on its competitors' results. It is necessary and natural that, before taking any initiative, the firm's manager should anticipate the reactions of the competition and prepare the appropriate response. This theory refers to competition stimulated by the sharp reduction in customs protection and by the free movement of capital. Exports are directly favourable to domestic employment and to imports of sales and production subsidiaries abroad, and they are the vectors of growth and competition. In addition, competition in these sectors, whether traditional or new, has more or less destabilising effects on the domestic market[36] .
The Duopoly theory raises other concerns that contradict Joan Violet Robinson's point of view. The latter states that competition is an unattainable situation in practice. In his theory,

[33] J. Boussard et al. 2005, *Libéraliser l'agriculture mondiale, Théories, modèles et réalités,* Paris, Montpellier, p.135.
[34] J. Berthelot, 2001, *La mise en boites du soutien interne ; Attention aux étiquettes mensongères*, Paris, Montpellier, pp. 218-246.
[35] M. Duopole, 1999, *La concurrence selon Porter, le village mondial*, Paris, Gallimard, p. 24.
[36] J. Kraft, 1999, *Le processus de concurrence,* Economica, Paris, Montpellier, p. 4.

Robinson contrasts imperfect competition with monopolies, considered by neo-classical theorists as a major case that corresponds to real economic life. For him, demand was the only source of expansion that could eliminate savings and reduce the role of perfect competition. He believes that an economy can prosper in a situation of imperfect competition. But the problem lies in the capacity of national production to absorb domestic demand[37] .

Adam Smith, on the other hand, was interested in international trade as a source of gain for nations, and at the same time opposed the mercantilists by putting forward two arguments: The first argument is that of absolute advantage. The first argument is that of absolute advantage. Imports generate a gain abroad[38] and it is advisable to buy abroad what is available at a high cost. Similarly, the domestic economy exports the goods for which it produces under more advantageous conditions. The second argument concerns the size of markets. Adam Smith's labour principle refers to an engine of economic growth limited by the size of the market. This means opening up the economy so that it can participate in a larger market and benefit from more efficient techniques. Smith demonstrates that all modern theories of international trade are linked to this idea of growth by referring to "international economies of scale[39] ".

With regard to this study, the Duopoly theory enabled us to approach part of our work, but it was much more focused on market competition without dealing with the aspect relating to household demand, which is what our work seeks to analyse.

VI - Chronological framework

The first European settlers in Africa were mostly missionaries and traders. Speaking of the chronological limits of this work, we have to justify two milestones ranging from 1928 to 2016.

The upstream milestone corresponds to the date 1928, which marks the year when cotton growing was introduced into Chad in general and the Mandoul region in particular, made compulsory by the colonial administration. On 05 May 1928, the colonial administration extended cotton growing to the south of the country and the development of ginning stations was made possible and fixed by the conventions instituting the policy of zones to be favoured for the cotton purchasing monopoly, following the example of the policy already practised in the Belgian Congo.

The downstream milestone, which corresponds to 2016, marks the launch of the agricultural campaign by the President of the Republic of Chad, with the emphasis on the new five-year term and the pursuit of mechanisation of agriculture to make it more intensive, and crop diversification. In his speech, he also planned to strengthen cotton production and marketing capacities. He appealed to international cooperation and diversified partnerships to invest in agriculture in order to boost production and guarantee food security for all Chadians. To this end, the Chadian government is encouraging cotton growers by increasing the price of cotton to 240 francs per kilogramme for the 2013/2014 and 2015/2016 crop years. The Chadian government is therefore urging farmers to invest in agriculture in order to boost the national economy.

---

[37] J. Robinson, 1993, *L'économie de la concurrence imparfaite*, Paris, Montpellier, p. 65.

[38] E. Hechscher, 1972, *Echange international et croissant, Economica*, Paris, PUF, p. 12.

[39] A. Smith, 1776, The Wealth of the Nation, http://www.leconomiste.eu/descryptage-economie/211-idee- clef-de-la-richesse-des-nations-d-adam-smith.html, accessed, 08 May 2017.

VII - Geographical context

The Mandoul Oriental region is located in the extreme south of Chad between latitudes 8 and 9 degrees north and longitudes 17 and 18 degrees east. It is bordered to the north by the Tandjilé and Baguirmi regions, to the south by the CAR, to the west by the Logone Oriental region and to the east by the Moyen-Chari region. It covers an area of 17,727km2 and has an estimated population of 558,416[40] according to the 2009 census.

With regard to its administrative situation, it is subdivided into three (03) departments, namely Mandoul Oriental (chief town Koumra) with six (06) sub-prefectures and nine (09) cantons, Mandoul Occidental (chief town Bédjondo) with four (04) sub-prefectures and six (06) cantons and Barh Sara (chief town Moïssala) with five (05) sub-prefectures and seventeen (17) cantons.

Overall, the Mandoul Oriental region has more than 1,125 villages, which are constantly being subdivided. The main ethnic and linguistic groups are: the Sara Madjingaye, the Gor, the Daye, the Goulei, the Nangnda, the Nar, the Mbaye and the Toumack.

However, the socio-cultural life of the population of East Mandul is animated by traditional and religious structures, rites and customs[41] . Today, this region is inhabited by a cosmopolitan population with several socio-cultural groups from different backgrounds. The traditional structures are influential and wield considerable power in making decisions affecting the lives of the local population. Most district chiefs are the sons or grandsons of the original occupants of the outlying areas[42] . The administrative structures are subdivided into several posts as mentioned above.

As far as religious structures are concerned, three main religions are practised in the East Mandul region: Christianity (Catholicism and Protestantism), Islam and Animism. It should also be noted that animism is only practised by a tiny proportion of the population. Christianity comes first, followed by Islam, but the lack of reliable statistical data makes it difficult to confirm this ranking for each of these religions. The rites and customs often practised are diverse and range from initiation ceremonies (yondoh, bayan) to traditional cults of imploring the dead "Nan Begue, Nan Sewe" to ask for their blessing on various occasions. These include the purification of villages, abundant rainfall and harvests. Belam" is the sacrifice during which followers implore the blessing of the ancestors for social well-being. Nan Sewe is a sacrifice to the dead to thank them for their protection in any form during the past year, and is performed at the end of the year. Nan Begue, the village purification sacrifice, is performed when an unfortunate event occurs in the village[43] .

The climate in the Mandoul Oriental region is Sudanian, with two main seasons: a dry season from November to April and a rainy season from May to October. Average annual rainfall varies between 600 and 1,200mm[44] , with a strong rainy period between July and September. Ferruginous soils cover almost all of the region[45] . Ferralitic soils are red in colour and highly sensitive to water erosion. The soil texture is sandy or silty-sandy[46] . The region also boasts a diverse range of domestic fauna, including cattle, sheep, goats, pigs and poultry. With the

---

[40] Rapport de la sous-préfecture de Koumra, 2009, Recensement de la population du Mandoul Oriental, p.15.
[41]Koumra Commune report, 2014, Communal development plan, p. 21.
[42] *Ibid.*
Report by the Commune of Koumra, 2014, pp.42-43.
ONDR report, 2012, La statistique sur la pluviométrie dans le Mandoul Oriental, p. 5.
Ibid, p.15.
Ibid.

exception of poultry, which are more or less kept on concessions, the rest of the domestic animals (cattle and pigs) are constantly on the move and cause serious agricultural damage. As for the wild fauna, it is poor and consists of a few rare reptiles such as margouillats, snakes and monitor lizards. The map below shows the Mandoul Oriental region among many others in southern Chad.

Figure 1: Location of the East Mandul region

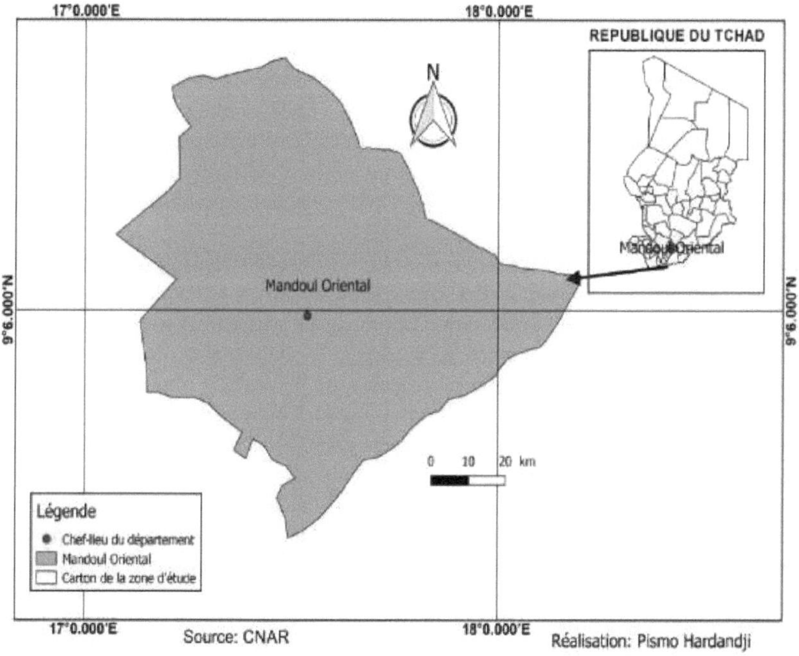

VIII - Literature review

The issue of production and marketing comes down to agricultural and commercial activities, which have given rise to numerous studies carried out by authors around the world. These studies focus on agriculture, considered to be one of the most important activities for global and national economic development.

Mahamat Sorto[47] at the Spirulina conference in Niger in 2006 and the Spirulina conference in Madagascar in 2008, spoke about a project to improve the production of traditional natural Spirulina, the "dihé" kannembou from the shores of Lake Chad. This project

[47] Mahamat Sorto, 2006, L'amélioration de la qualité de Dihe, la spiruline récoltée au Tchad, Spirulina-gadez.free.fr/0503pm.htm, consulted on 05 March 2017.

project began in 2007, when production of traditional dihe was reported to be in the region of four hundred tonnes. Nine very active women's groups, supported by the project, produced around ten tonnes of improved dihe in 2007 and 2009. With a few additional improvements, export is envisaged in the near future at a cost price of around 8 euros per kilogram[48] . The project is funded by the European Union and managed by the United Nations Food and Agriculture Organization (FAO).

In 2010, a film was made by Mahamat Sorto and the women's groups showing the improvements made to the dihe harvest: sieving of the biomass, filtration on cloth, hand-pressing, extrusion and solar drying under shelter, grinding at the mill and packaging in heat-sealed plastic bags. The author wanted to bring a new, modern approach to the production of dihe, which used to be made in the traditional way.[49] This work enables us to analyse the content of our subject on the production and marketing of cotton, but without taking into account the policy of the states concerned in terms of the price of dihe on the external market. To this end, our work highlights the intervention of states in the production and management of cash crops in the franc zone.

Antoine de Montchrétien[50] in the history of economic thought states that agriculture, industry and commerce are interdependent. For him, agriculture plays a key role in the economy: "ploughing [...] must be considered as the beginning of all faculties and wealth"[51] . But if agriculture is fundamental, it must be accompanied by industrial production, which is the future and the complement of agricultural products. As for trade, it is an essential element in the expansion of a society based on the division of labour and the pursuit of technical progress. A. de Montchrétien points out that needs generate demand, which drives production. It is the quest for profit that determines human action.

Baba Dioum et al[52] have shown that in Africa, as elsewhere in the developing world, agricultural development contributes more to economic development than any other area. This has led to an overall increase in incomes in rural areas, where the majority of vulnerable and poor people live and work. In this way, the agricultural sector stimulates the development of other sectors of the economy, including demand for goods and services produced outside the sector. In addition, this sector positively reduces the level of poverty, famine and malnutrition by increasing the food supply. It also improves access to better food through higher incomes in rural areas and in other sectors of the economy[53] . This work has enabled us to study the dynamics of agriculture in Africa, but ignores its negative impacts on the environment, which our work aims to address.

Beyem Roné[54] shows the role of cotton growing in the industrialisation of southern Chad[55] .

---

[48] Mahamat Sorto, 2006, p. 7.

[49] http://www.dailymotion.com/video/Xcu89hproduction of - spiruline-au-Lac-tchao lifestyle, accessed on05 March 2017.

A. de Montchrétien, 1993, *Histoire des pensées économiques*, Paris, Sirey, p.89.

Ibid.

[52] B. Dioum et al, 2008, "Cadre pour l'amélioration des infrastructures rurales et la capacité pour l'accès au marché, étude réalisée par un groupe d'experts, sous les hospices de la conférence des ministres de l'agriculture de l'Afrique de l'Ouest et du Centre (CMA/AOC)", pp. 17-26.

[53] ECA, 2012, "Regional integration in West Africa: regional agricultural value chains to integrate and transform the agricultural sector, Economic Commission for Africa. Sub-Regional Office for West Africa", p. 32.

[54] Beyem Roné, 2000, Tchad : *l'ambivalence culturelle et l'intégration nationale*, Paris, L'Harmattan, p. 82.

[55] J-L. Charléard, 2003, *Cultures vivrières et commerciales en Afrique Occidentale*, Nantes (France), Éditions du temps, pp. 267-447.

In his view, cotton growing predates the Ngarta Tombal- baye regime, dating back to the colonial period. The installation of industries obeyed economic rules relating to profitability and the existence of raw materials where an industry was to be installed. The author mentions the Mayo-Kibbi region, which did not benefit from certain industries even though it met the necessary conditions to welcome colonial administrators. This work enables us to study the diplomatic relationship between Chad and the countries of the North with regard to cotton growing. However, the concerns of the local population have not been taken into account, particularly with regard to the impact of cotton growing, which we are interested in analysing in this study.

Robert Buijtenhuijs[56] addresses the issue of Chad's economic exploitation (cotton growing) as having contributed to a regional disparity accentuating the North-South divide. His study focuses much more on the political dimension of the subject, describing the role played by the cotton economy in the face of the country's crises. In his view, the introduction of cotton cultivation led to a certain upheaval in the structures of traditional societies. He believes that the main beneficiaries of cotton are COTONFRAN, which later became COTONT-CHAD, C.F.D.T (Compagnie Française de Développement des Fibres Textiles) and the Chadian government, but not the cotton growers. We find this work interesting insofar as the author stresses the importance of cotton growing for the benefit of the Chadian state and the companies in place, but fails to take account of other crops that can help farmers on the same land where they grow cotton. It is the effects of fertiliser on the cultivable soil that we are interested in studying.

Jean Pierre Magnant[57] discusses the mechanism of the imposition of cotton cultivation after recalling the aim of colonisation, which is the economic exploitation of the colonies. He also looks at the question of the capitation tax and other duties to which the populations were subject. However, Magnant focused his thoughts particularly on the eastern part of the south of the country.

Régine Levrat[58] , published in 1950, describes the population movements organised by the Cameroon government as the main factor in the development of the far north. Not only did they relieve the pressure on the northernmost areas, where the level of natural resources was no longer sufficient to meet the needs of the population, but they also enabled the development of other, more southerly, climatically favoured areas. He has shown that cotton production is not an end in itself, especially in a country where the textile industry is virtually non-existent, and if Cameroon's production were to disappear tomorrow, this would not move the indices one hundredth of a point. Levrat explains that cotton from North Cameroon, in accordance with the wishes of its promoters, has really played the role of "locomotive" of development that was assigned to it. It is the main source of income for farmers, and has brought them into the national economy. In addition, he points out that cotton cultivation favours cereal crops which
have enabled the cotton-growing zone to achieve food self-sufficiency[59] . In addition to cotton production, most of the fibre is exported, generating significant foreign exchange earnings,

---

[56] R. Buijtenhuijs, 1978, *Le Frolinat et les révoltes populaires du Tchad*, Paris, Mouton, p. 78.

[57] J. P. Magnant, 1987, *La terre sara, terre tchadienne*, Paris, L'Harmattan, quoted by ARMI Jonas, 2003, "L'économie cotonnière dans la région de Pala au Tchad (1925-2000)", Master's thesis in History, University of Ngaoundéré, p. 5.

[58] R. Levrat, 1950, *Culture commerciale et développement rural, l'exemple du coton au Nord-Cameroun,* Paris, L'Harmattan, pp. 5-6.

[59] R. Levrat, 1950, p.11.

table oil is an important part of the household basket, and livestock feed has become essential for livestock farmers[60] .

In view of the above, a great deal of work has been carried out in the agro-industrial field, more specifically in the production of cotton and other cash crops throughout the world, as well as food crops. However, this work is limited to the scientific production of cotton, and this is the case of our present study in the Mandoul Oriental locality (Koumra in Chad).

IX - The issue

The introduction of cotton growing in Chad and the Mandoul Oriental was aimed at exploiting the local population through forced labour and capitation taxes. This crop has been a real lever for the country's development, but in recent years it has been hit by the ups and downs of the international market, causing a drop in cotton production. Cotton is a development tool for many African countries through trade policies. On the eve of the Sixth WTO Ministerial Conference, it is clear that the instability of the African cotton sectors in general, and that of Chad in particular, remains a cause for concern and a brake on development[61] .

Cotton production poses the problem of the development of the Mandoul Oriental region as it evolves. To this end, the following question arises: How has cotton production and marketing developed in the Mandoul Oriental region of Chad, and what impact has it had? This main question gives rise to the following secondary questions: Was cotton growing the cause of Chad's opening up to the outside world? Is it the direct cause of the modernisation of the agricultural system in Chad in general and in Mandoul in particular?

X - Research objectives

As part of this study, we have identified one main objective.

This objective is to show the dynamics of cotton production and marketing in the Mandoul Oriental region of Chad. This main objective is broken down into specific objectives.

- Specific objectives

> To show the background to the introduction of cotton in Chad in general and in the Mandoul Oriental (Koumra) in particular;

> Presenting cotton production;

> Study the marketing of cotton ;

> To demonstrate the impact of cotton production and marketing on the socio-economic development of the Mandoul Oriental region.

XI - Methodology

As part of this study, we used a two-stage methodology. The first consisted of collecting written documents such as books, articles, theses, dissertations, internship reports and digital documents. The second stage involves collecting oral sources. The latter plays an important role in the context of African historiography. Oral sources are materials recorded during interviews with witnesses. It is much more highly regarded by the pioneers of African history, and taught in African historical schools: the Ibadan Historical School, the Dar Es-Salam Historical School and the Dakar Historical School[62] . Joseph Ki-Zerbo said that being a historian "means choosing your subject, your centres of documentation and your sources[62] ".

---

60 *Ibid.*

61 E. Hazard, 2005, *Enda prospectives dialogues politiques*, enda éditions, Dakar, p.103.

62 J. Ki-Zerbo, 1980, *Histoire générale de l'Afrique Tomme I*, Méthodologie et Préhistoire Africaine, p.

It is with this in mind that oral sources are valued in African history. In addition, there are two important aspects to this methodology: the method and the techniques. These two elements incorporate a multidisciplinary approach, with recourse to other auxiliary disciplines such as sociology, geography, anthropology and law.

After collecting the data, we processed it. During the research, we visited several documentation centres: the Library of the Faculty of Arts, Letters and Human Sciences (FALSH) of the University of Ngaoundéré, the Central Library of the University of Ngaoundéré, the Archives of COTON TCHAD-SN in Koumra, the Archives of the ONDR, the Archives of the Ministry of Agriculture and Rural Development, the Archives of the Koumra sub-prefecture, the Archives of the Koumra Commune, the Archives of the Bureau d'Étude et de Liaison d'Actions Caritatives et de Développement (BELACD) and the Archives of the Centre d'Étude et de Formation pour le Développement (CEFOD). These various documentation centres enabled us to carry out this work. Once the data had been collected, we had to analyse and classify them. It is important to note the iconographic sources produced using digital cameras, which constitute the images in this work.

XII - Research difficulties

This study focuses on cotton production and marketing in the Mandoul Oriental region (Chad) from 1928 to 2016. We encountered difficulties in relation to the distance from the study area, which meant that we had to have the financial means to travel. We had difficulty accessing certain documents for various reasons: the lack of suitable premises for storing documents, the lack of qualified staff to classify documents, and the unrest in the country, which led to the burning of archives and the looting of documents in the various departments. We have difficulties collecting oral data due to the mistrust of certain staff during the survey, and we are stuck with the period of field work when we have to follow the farmers to the field in the rain in order to collect information.

XIII - Work plan

The work plan allows researchers to move through the research by chapter.

The first chapter describes the context in which the cotton sector was established in Chad in general and in the Mandoul Oriental (Koumra) in particular. The aim is to show the context in which cotton growing was introduced in southern Chad.

The second chapter presents cotton production in the Man- doul Oriental region. It deals with the factors and systems of cotton production through its extension.

The third chapter analyses cotton marketing at farm level and at the Koumra ginning plant. The aim of this chapter is to show the process from the purchase of seed cotton to its transformation into a finished product that is traded both internally and externally.

The fourth chapter looks at the impact of cotton production and marketing on socio-economic development in the Man- doul Oriental region. The advantages and disadvantages of cotton production for socio-economic development are examined.

CHAPTER I

THE BACKGROUND TO THE ESTABLISHMENT OF THE COTTON INDUSTRY AUTCHAD

Chad is a Central African country where cotton cultivation plays a major role in the country's agricultural fabric. As well as generating substantial revenue for the State, cotton provides a livelihood for more than 3 million Chadians and is a real lever for rural development, particularly in the southern part of the country[63] , a cotton-growing area par excellence. As a landlocked Sahelian country, the Republic of Chad has significant agro-sylvo-pastoral potential[64] . Agriculture and livestock farming remain the only means of subsistence in rural areas. One of the main cash crops (along with groundnuts), cotton growing was imposed in southern Chad from 1920 onwards, a region commonly referred to as "useful Chad"[65] . The forced cultivation of cotton introduced by the French colonists destabilised farming systems and the organisation of societies. Its adoption led to the almost total abandonment of cereal growing and a consequent impoverishment of the soil. As a result, this crop has often been equated with food insecurity, as it monopolises the entire peasant workforce to the detriment of subsistence farming. Nonetheless, cotton production in Chad is on the rise, even if it has been hit by a number of crises in recent years.

I - HISTORY OF COTTON GROWING IN CHAD

The idea of developing cotton growing in Chad dates back to colonial times and was designed to serve the economic interests of the coloniser. At the time, the French textile industry was dependent on cotton imports from the United States and the French colonies in Asia. At the time of the American Civil War, France was experiencing difficulties in supplying its textile industry with raw materials. To avoid bankrupting France's most important manufacturing industry, metropolitan France initiated a policy of developing cotton growing in its African colonies, starting with Afrique Occidentale Française (AOF) and then Afrique Équatoriale Française (AEF)[66] . Territories or zones where cotton growing was established were defined in AEF, more specifically in Chad, Ubangi Chari and later in Cameroon. This is recorded for the first time in the

Lenfant's 1904 mission to Chad was characterised by a conquered census. He wondered about the indifference of the indigenous population towards the wild cotton plant[67] .

In 1910, a military commander in Chad (Colonel Moll) drew up an embryonic "*cotton plan*"; his work could not be continued because of the 1911 territorial agreement, which ceded south-west Chad to Germany[68] . It was not until the aftermath of the First World War that cotton-growing trials were resumed. The first official trials of cotton cultivation in Chad were conducted in 1921 by Captain Delinguette. He cultivated 500 hectares of cotton on the shores

[63] C. Arditi, 1999, "Paysans Sara et éleveurs arabes dans le sud du Tchad", *in L'homme et l'animal dans le bassin du Lac Tchad. Editions IRD, collection colloques et séminaires,* pp. 557-575.

[64] G. Rameau, 1953, "L'élevage bovin au Tchad", *Revue internationale des produits coloniaux et du matériel, number 282,* C.A.O.M., p. 332.

[65] G. Magrin, 2001, " Le Sud du Tchad en mutation " : *des champs du coton aux sirènes de l'or noir,* Doctoral thesis in Geography, University of Paris, Panthéon-Sorbonne, pp. 102-113.

[66] Antonetti, 1927-1931, Speech and report on the general situation of the A.E.F., p. 68.

[67] Lenfant, 1909, " Le coton pousse à vue d'œil à proximité des cases, par l'apathie de l'indigène, reste inutile et se perd ", *La découverte des Grandes sources du centre de l'Afrique*, Vol. 37, N° 54, p. 25.

[68] *Ibid.*

of Lake Léré and the Mayo-Kebbi, producing a yield of one tonne per hectare[69] .
Industrial cotton growing was finally introduced in Chad in 1928. Initially imposed as a compulsory and forced crop, it gradually gained the support of the populations in the cotton-growing area. Since the 1980s, cotton has been claimed and accepted by the populations of the south of the country, due to its importance and socio-economic interest at various levels, and its strategic position for Chad. However, in the Mayo-Kebbi West region in south-western Chad and northern Cameroon, indigenous cotton was known as a pick-your-own product and a backyard crop, before industrial cultivation was introduced. The local populations picked cotton, ginned and spun it by hand, then wove it on handicraft looms into narrow strips of fabric called "gaback" used to make garments commonly known as cover-ups[70] .

A. Cotton-growing policy during the colonial period

1. Involvement of the administration in sharing out the cotton production zone

As early as 1928, the colonial administration granted the Compagnie Cotonnière Congolaise (COTONCO), founded in Brazzaville in 1926, a monopoly on the exploitation and marketing of Chadian cotton under a contract signed with the company. French policy was characterised by the theory of "association" and an awareness of its own "moral superiority"[71] . Six (6) years later, four cotton companies had been founded in Africa, and agreements had been drawn up allocating roles between the concessionary companies and the colonial administration. One of these companies, the Compagnie Cotonnière de l'Afrique Équatoriale Française (COTONFRAN), an offshoot of COTONCO, was granted the entire cotton zone of Chad as well as four subdivisions in the north of the Central African Republic. COTONFRAN has a monopoly on the purchase of production, with an obligation to buy at least 80% of the harvest[72] . In addition, the agreements of 05 May 1928 set out the obligations of the cotton companies, which had to undertake to build ginning plants. The purchase price of cotton from producers would be set each year on the basis of market prices in Le Havre[73] . When the agreements were renewed in 1939 and 1949, the content of the original contracts was maintained.
The administration introduced a number of new obligations, such as reducing the portage involved in transporting cotton to the factory, and setting up purchasing centres in the villages. From 1934, four companies were divided up: COTONFRAN received the cotton lands of Chad, plus the subdivisions of Bouali, Fort-Sibut, Dekoa and Batangafo in Oubangui; COTONAF, COMOUNA and COTONBANGUI each received the privilege of purchasing in the Oubangui regions. What these companies have in common is that they are mainly foreign-owned (Belgian and Dutch). However, their boards of directors included French personalities who had long been associated with colonial circles. Indeed, before 1930, each company had set up an experimental centre in its own territory. For example, COTONFRAN initially set up two small factories in Moissala and Doba in southern Chad, comprising two gins with 40 motor saws and a seed selection and improvement farm at Bekamba. Two other high-capacity

[69] W. Stevelinck, 1953, "Le développement du coton dans la zone Mayo Kebbi, Logone et Moyen Chari", *Marches coloniaux du monde,* number 399, C.A.O.M., p. 75.
[70] Prepared by COTONTCHAD SN and Ministry of Agriculture, p. 25.
[71] A. Sarraut, 1932, *La mise en valeur des colonies françaises*, Paris, Payot, p. 103.
[72] Office National de Développement Rural, 1977-1981, *Proposition pour un programme dans la zone Sud du Tchad,* p. 45.
[73] J. Cabot, La culture du coton au Tchad, Annales de Géographie, Vol.66, N°358, pp. 99-104.

factories, each with two 60-saw gins, were then set up at Koumra and Lai[74] . In addition, as part of the agreement signed with the cotton companies, in 1932 the Governor of A.E.F. Anttonetti set up a semi-public agronomy department at Fort-Archambault, headed by an agricultural engineer. A year later, two main stations had been set up, one in the centre of Fort-Archambault (now the town of Sarh) and the other near Grimari in Oubangui-Chari. In the same year, a substation was installed at Bebedjia in Chad[75] . From then on, scientific studies were carried out to test cotton growing in Africa.

2. Private initiatives and the involvement of the colonial authorities in the experimental phases of cotton production

At the beginning of the 20th century, on the initiative of private bodies dominated by the Lobbies of the European cotton industries, trials of cotton growing were undertaken in most regions of Africa. The movement to promote cotton growing on the dark continent had started in Germany, where the KOLONEL Wirtschafliches komitee had carefully conducted an experiment in Togo as early as May 1900. In addition, from January to July 1901, he entrusted the American J.C. Calloway to study the possibility of growing cotton in a German sphere in Africa. He carried out scientific trials in the Misahohe district on the Guinea coast. However, the results of these trials were inconclusive due to the humidity in the region. However, the 1902-1903 campaign enabled 16 bales to be shipped to Bremen. During the same period, trials were carried out in East Africa in the Kiloua, Pongwe and Tanga districts, but without much success. From then on, the territories of Chad and especially the Mayo-Kebbi region, under the French flag at the time, appeared to Germany to be the cotton-growing countries of the future[76] . It was in this same spirit that the British Cotton Growing Association was founded in Manchester in 1902, with the aim of promoting cotton growing in the English colonies in Africa. At the beginning of 1903, a mission of six experts was sent to the various English colonies in West Africa to study the possibility of cotton production from the point of view of the varieties to be used, labour and transport. The various trials were promising, especially the one carried out around Ibadan and Abeokuta on 2,400 hectares at the instigation of the Governor of Lagos, Sir William Mc Gregor, which resulted in 70 tonnes being landed in England. The cotton harvested was of good quality and the British Cotton Growing Association had great confidence in the future of cotton in the Nigerian territories, particularly in Ka- no[77] .

However, inspired by these models, like the English association, in January 1903 the Colonial Cotton Association was founded in Paris under the aegis of Albert Esnault-Pelterie, President of the Syndicat Général de l'industrie cotonnière française. Its aim was to develop cotton growing in the French colonies and to promote the use by French industry of the raw material harvested[78] . The first attempts at intensive cotton growing date back to 1903 in the middle Niger valley in West Africa. They had to be abandoned around 1909 due to repeated failures[79]

[74] C. Marquet, 1930, *L'orientation économique et financière, administration générale*, CEFOD-TCHAD, reference CFB00523, p. 45.

[75] G. Boussenoul, 1939, La culture cotonnière en A.E.F. *la revue politique et parlementaire,* pp. 86-99, quoted by ABAKAR KASSAMBA Abdoulaye, 2010, *La situation économique et sociale du Tchad de 1900 à 1960*, doctoral thesis at the University of Strasbourg, p. 242.

[76] M. Zimmermann, 1911, "L'accord franco-allemand au sujet du Maroc et du Congo", *Annales de Géo- graphié, Année* 1912, Volume 21, Numéro 116, pp. 88-91.

[77] G. Boussenoul, 1939, p. 132.

[78] Secretary of State for External Affairs, Economy and Development Plan, B.D.I.C, September 1968, p. 27.

[79] Ministère de la finance d'Outre-mer, A.E.F, Tchad, Paris, 1948, B.D.I.C, p. 13.

.

On the other hand, the German colonies in East Africa supplied 551 tonnes of cotton in the same year, despite the unfavourable conditions in these regions for this crop[80] . At the same time, in the north of French equatorial Africa, similar steps were taken to promote cotton growing. For example, before the cotton association was founded, the French cotton union entrusted Doctor Decourse with a mission to Chari-Lac Tchad from 1902 to 1904, when samples of fabrics were submitted to merchants in Fort-Lamy (now N'Djamena) and local tailors. He was able to classify the most sought-after fabrics in Central Africa and provide an overview of the prices offered by local merchants[81] . It was with this in mind that Captain Lenfant was given a mission to the Bénoué-Tchad region from 1903 to 1904 to look into the question of supplying Chad and also to study the cotton prospects of the Chad colony. However, the territory of Chad could provide an interesting raw material for Metropolitan France and a main activity for most of the Chadian population, whose interests extended beyond the cotton-growing area[82] .

3. The aborted attempt to introduce cotton to the regions of Chad

Following the success of his mission to demarcate the border of the colony of Chad with German Cameroon, Major Henri Moll was given command of the military territory of Chad in 1909. As soon as he arrived, he set up a cotton production plan in the Mayo-Kebbi region, where he foresaw a more promising future for this crop[83] . However, his death in November 1911, which led to an agreement to grant major territorial concessions in East Africa to Germany in exchange for France's supremacy in Morocco, put an end to this programme. The regions of Chad granted to Germany under this agreement included the Logone Oriental and the Mayo-Kebbi with Léré. Through this agreement, Germany had gained access to vast regions promising for cotton cultivation, a crop in which it had never concealed its interest[84] . However, the Germans had carried out several trials to develop cotton growing in the south of Cameroon, but the experiment was abandoned because of the humidity in the region. They then shifted their efforts to the north, where a modest hybrid experimentation station for food crops and cotton had been set up at Kousseri and another at Pitoa near Garoua, close to the confluence of the Benoué and Mayo-Kebbi rivers, created in 1912 by Dr Wolf with 450 hectares devoted to cotton[85] .

B. The economic and social context of cotton growing in Chad during the colonial period

Cotton growing has played an undeniable role in Chad's economic and social development since it was first introduced to the country.

1. The role of cotton in the economy

There are various reasons why the authorities and the State encouraged cotton growing. When cotton growing was first introduced, financial prospects played an important role. On the one hand, export tax revenues were expected; on the other, cotton seemed to facilitate the introduction of the capitation tax in monetary form, which was levied by the colony. Before

---

[80] M. Zimmermann, 1911, pp. 185-188.

[81] M. Zimmermann, 1911. 189.

[82] A. P. Eya'a Akoumba, 2002, Organiser la production et la commercialisation du coton dans la zone soudanienne, file: //D/ Restructuration de la filière coton au Tchad. htm, accessed 17 August 2017.

[83] A. Chevalier, P. Senay, 1949, *Le coton*, Paris, PUF, p. 223.

[84] *Ibid.*

[85] R. Levrat, 1950, *Le coton en Afrique Occidentale et Centrale avant 1950, un exemple de la politique coloniale de la France*, Études africaines, Paris, L'Harmattan, p. 188.

the Second World War, the tax on cotton exports was abandoned because of the difficult situation on the world market and, in the 1930s, farmers' income from cotton was often lower than the yield from the head tax[86] . Independent Chad also saw cotton as an important source of income. Cotton was overestimated as an export product and supplier of foreign currency. As cotton's share of exports remained the same, the possibility of covering imports with cotton exports fell to less than 40%[87] .

In addition, one principle has prevailed in cotton marketing: whatever the ups and downs of production and fluctuations in world market prices, the companies are guaranteed their operating costs. As a result, during the massive call for textiles that followed the end of the war, the companies' profit margins increased significantly, without any corresponding improvement in farmers' remuneration. A cotton support fund, set up to balance the companies' income, received a share of the profits made when world cotton prices rose to cover their costs and the payment of their dividends[88] . As for the remuneration of local producers, this is left to the discretion of the authorities. Yet the companies claim to be the only profitable economic activity in the country. A concerted effort has therefore been made to improve yields per hectare. This requires strict adherence to sowing dates, which have the disadvantage of coinciding with the sowing period for food crops[89] . Understandably, the farmers' first priority is to prepare the millet fields, whose early harvest will whet their appetites. This choice automatically leads to a drop in cotton yield per hectare.

Until 1952, the Cameroonian border regions surrounding the district produced cotton but did not grow any industrial crops[90] . The Foul- bé of Binder did not hesitate to grow their cotton on the hut fields with soils enriched by manure, and to relegate their millet and pea fields to the old cotton soils. Yields reached and exceeded 500kg of seed cotton per hectare. However, since the Compagnie Française des Textiles organised cotton growing, millet harvests have dropped significantly and the Foulbé of Binder have seen the price of millet rise rapidly on their markets when millet is available for sale[91] . They had to revert to alternating millet and cotton in the hut fields. Among the Massa, who live to the north of Bongor, cotton growing is practised with great reserve. Young Massa find it more profitable to engage in seasonal fishing down the Logone between Katoa and Fort-Lamy[92] .

2. Social issues

The relationship between social cohesion and the level of production, and the social structure have an impact on productivity: villages whose inhabitants are united by family ties, affinities or a community of geographical origin have a higher level of production. This is because farmers are united in competition: cotton production is a matter of honour and social status, both within the village and between villages. On the other hand, mixed villages where

[86] L. Sanz, 2015, La route du coton en Afrique Equatoriale Française, file//D/Jo./ La route du coton en AEF (voyage-Congo.over-blog.com) - DMCARC.htm, accessed on 19 July 2017.

[87] *Ibid.* p. 34.

[88] *Ibid.*

[89] S. Marco, 1985, "Le problème des cultures obligatoires dans les produits d'exportation", *Histoire générale de l'Afrique VII. L'Afrique sous coloniale 1885-1935,* Paris, Unesco, pp. 105-109.

[90] G. Magrin, 2001, p. 109.

[91] E.D. William, 1910, "Le coton, sa culture, son commerce", *La quinzaine coloniale, fourteenth year,* www.persee.fr/doc/ingeo_0020_0039_1956_rum_20_2_1571, accessed 30 July 2017

[92] A survey by M. Blache, a hydrobiologist with the O.R.S.T.O.M., revealed that, in a group observed at the confluence of the Logone and Chari rivers, each fisherman earned between 75,000 and 55,000 FCFA in three months. See also Jean Cabot, populations du Moyen Logone, l'Homme d'Outre-mer, n°1,1955, Paris, O.R.S.T.O.M, p.13.

descendants of the original occupants of the land, known as "children of the village", live side by side with recent emigrants, generally, but certainly not always, experience lower productivity"[93] . The link between productivity and social structure can be seen in the way village cotton groups operate.

Producers do not negotiate individually with cotton buyers, but organise themselves into village groups to deal with the cotton company collectively. Where these associations more or less follow the configuration of existing groups, they operate efficiently and very productively. But where there are no pre-established social links, these "associations are weakened by a lack of legitimacy and accountability"[94] . Ultimately, the degree of social cohesion that exists between group members influences, and is characterised by, practices of hoarding and fraud.

Overall, there is a greater degree of social cohesion in the western regions of the cotton-growing zone than in the eastern and central regions. In the western regions, where the ethnic majority is Moundang, and in the central region (among certain sub-groups of the Sara ethnic group), the villages are characterised by a highly structured and hierarchical socio-political organisation[95] . The local authorities are considered legitimate and the elders are respected for their ancestral knowledge.

The income from cotton cultivation is used to purchase community and development facilities that are also accessible to non-producers. In fact, cotton itself is seen as a public asset for which the whole community is responsible. In the east, however, as in most of southern Chad, the social and political structures of power are less legitimate and less cohesive. Although the various ethnic groups coexist peacefully and ethnic, linguistic or cultural differences are minimised, cotton production causes rifts and high social costs: violent clashes between members of the same family and with administrators, assassinations, exiles and expulsions from villages[96] .

11- COTTON GROWING DURING THE COLONIAL PERIOD

A. Cotton production under COTONFRAN in Chad

To ensure the production and marketing of cotton, the colonial administration set up a fairly coherent structure. There was a clear division of tasks between the various players in the cotton industry: first there was the colonial administration, then the Compagnie Française du Développement des Fibres Textiles (CFDT) and COTONFRAN[97] . Each plays a specific role in the production and marketing of cotton.

In terms of the cotton industry, the starting point is the cotton field to the buying centre, from the buying centre to the ginning plant, and from the ginning plant to the packaging and processing of the cotton fibre, which should be sold at world prices[98] . As mentioned above, the role of the colonial administration was to encourage the introduction of this crop. This was done by will or by force without taking into account the aspirations of the local populations[99] .

[93] Interview with Nassaryem, Koumra, 05 June 2017.

[94] Abakar Gouni Ousman, 2010, "Le commerce extérieur du Tchad de 1960 à nos jours", PhD thesis, University of Strasbourg, p. 76.

[95] *Ibid.*

[96] *Ibid.* p. 84.

[97] A. Chevalier and P. Senay, 1949, p. 223.

[98] E.D. William, 1910, p. 203.

[99] Abdoulaye Abakar Kassambara, 2010, "La situation économique et sociale du Tchad de 1900 à 1960", PhD

It was the will to implement this policy that led the colonial administration to have allies within the local populations (traditional chiefs and boycotton). This clearly shows the role played by the colonial administration in promoting cotton, which was seen as a key element in the task of agricultural extension. District chiefs were "the most active agents linking the upper hierarchy and the peasantry"[100] .

The second player in the system was the CFDT, which was a privileged partner of the colonial powers in the introduction of cotton growing in Chad. Indeed, it should be noted that the metropolitan textile industrial group threw all its weight behind France's decision to initiate this activity. The CFDT, which is in a sense an offshoot of this group, could not but be a stakeholder in this venture. It is a technical assistance company whose main role is to carry out agricultural modernisation in rural areas. This modernisation mainly concerned the popularisation of ploughing and the use of chemical fertilisers and plant protection products[101] .

COTONFRAN was a major partner in the cotton system, being at the other end of the chain. COTONFRAN is a mixed capital company with a monopoly on the purchase and processing of cotton into fibre. It was also required by the government to purchase all cotton production, and was omnipresent, both in the buying centres and in the factories. COTONFRAN also played an important role in propaganda in favour of cotton growing. The distribution of cotton rents to farmers by COTONFRAN was an opportunity to extol the merits of this progressive crop and to make farmers hope for a better future as long as they grew cotton[102] .

Given the role played by each of the companies mentioned, it is clear that the way in which cotton production and marketing were organised during the colonial period was designed solely to serve the interests of the colonial administration, to the detriment of the local population.

In the Mandoul Oriental region, farmers have been given the status of players in the cotton company. They have played an important role in cotton growing. Unfortunately, however, they have had to endure a lot of hard labour in the heat. The peasant farmer played a major role as a producer in the colonial cotton system, but in terms of spin-offs, he was marginalised.

1. Conditions for growing cotton

Cotton is grown on land that has been cleared, i.e. on newly cleared land. Sometimes cotton is grown on land left fallow. The choice of land to be cleared is usually made by the boy-coton (Agriculture Department official), accompanied by the head of the land or village[103] . The latter determine one or more plots to be cleared according to the number of cotton growers, taking into account whether or not the farmers have millet land.

The size of these plots is determined by the number of cords (cultivable areas) owed by the village, on the understanding that each peasant (aged between 15 and 50) must cultivate a field of cotton[104] . The physical pressure to grow cotton began very early on. Even during the trial years, before the contracts were signed with the cotton companies, it was common

thesis, University of Strasbourg, p. 109.

[100] Interview with Rimtoibaye, Koumra, 06 June 2017

[101] G. Diguimbaye and R. Langue, 1969, *L'essor du Tchad*, Paris, PUF, p. 30.

[102] G. Magrin, 2001, p. 63.

[103] M. Gaid, 1956, Au Tchad, les transformations subies par l'agriculture traditionnelle, notamment sous l'influence de la culture cotonnière. Comité de Coordination de la Recherche Agronomique et de la Production Agricole. Gvt. Général de l'A.E.F., p. 57.

[104] "Corde": unit of surface area used in Chad, a rope used to measure the side of a square to be cultivated.

practice to force farmers to work using the police. This violence originated not only with the colonial administration, which had undertaken to do so under the contracts with the cotton companies, but also with the local chiefs, who were keen to keep their bounty. The obligation to grow cotton was not officially abolished until 1956[105] . But the interest of the chiefs in this crop remained unchanged thanks to the premium, and the administration even continued to exert more or less strong pressure.

However, the chord is the standard measurement for the side of each individual field. It is generally 70 metres, which brings the compulsory crop to 49 ares per adult. In cotton-poor regions, the cord is sometimes as short as 60 metres (36-acre fields)[106] . Despite the involvement of local chiefs in cotton growing, the colonial authorities wanted to rely on traditional cultivation methods to increase cotton production and thus ensure tax collection. It should be noted that in southern Chad, decisions on sowing and harvesting were in the hands of the elders, who acted as land chiefs, religious chiefs, rain chiefs and bief chiefs. Most of these chiefs took advantage of their social position to reap individual benefits, as they had their constituents cultivate cotton fields from which they alone benefited. According to Jean Cabot, this system led to the resurrection of the "corvée seigneuriale" in the cotton-growing areas[107] . This practice was tolerated by the colonial administration during this period. It sought to interest local chiefs and make them dependent on cotton growing. Moreover, in order to get their money's worth, the chiefs were much more demanding of their subjects in order to increase yields. But this practice increased the disaffection of the village population with cotton growing. It was exacerbated by the fact that it ran counter to the traditional social foundations based on the equitable sharing of harvests that prevailed before the introduction of cotton. In fact, collective cultivation and the payment of cotton revenues to chiefs further discouraged farmers from growing cotton: "many mistakes had been made. Forced cultivation, payments to chiefs and collective cultivation were all used, contrary to the indigenous mentality[108] . Some of the chiefs who monopolised cotton cultivation tried to mitigate their unpopularity by paying the workers who came to their fields.

Some were content to provide food: meat and millet, while others only gave drink. Chiefs with the largest number of cords gave nothing or almost nothing, no doubt refusing to give this payment, which was very small in relation to their income, which they were required to disburse[109] . For this reason, the system further aggravated the chiefs' claims to benefit from the three-year cultivation system, by having the millet harvested on the rope for their benefit for a few years after the ownership of the ears of corn had grown, thus favouring the fallow period.

2. Forms of resistance and their impact on the population

The colonial administration's planned policy of exploiting the cotton industry and the economy came up against the reluctance and even the refusal of the people of the Mandoul Oriental region. They lived off subsistence farming, which was limited to activities on small plots of land, but ensured their food self-sufficiency. However, this cash crop was made compulsory, so people had no choice but to grow it at the expense of food crops. Given the

105 Interview with Tchadingar, koumra, 20 June 2017.

106 Abdoulaye Abakar Kassambara, 2010, "La situation économique et sociale du Tchad de 1900 à 1960", PhD thesis, University of Strasbourg, p. 124.

107 J. Cabot, 1957, "La culture du coton au Tchad", *Annales de géographie,* volume 66, number 358, p, 499-508, C.A.O.M.P, p. 149.

108 Governor General Reste, 1976, p.102.

109 P. Hugot, 1965, p. 143.

difficult conditions under which cotton was grown and the lack of interest shown by the population, some farmers decided to raise awareness among others in order to change the situation. It's worth noting the "exactions and humiliations suffered by the people, who were quick to respond to the colonial administration"[110] . As proof, the latter acted very badly against traditional values, with a lack of respect for the elderly, and for men who had been molested in front of their wives. For this reason, many of them used strategies to escape from the unscrupulous colonial agents.

In some villages in Mandoul Oriental, people fled to hide in the bush, because it was a fierce battle between them and these agents. People did not use the same strategies when confronted by the agents.

Some have sought to use more useful and peaceful means instead of attacking the oppressor directly. In fact, some farmers have migrated to be elsewhere, either by creating new villages or by going to countries bordering Chad. Peasants who have used peaceful means spend all their time in the bush and change their way of life in the absence of paternal love for their children. From then on, the instability of the Mandoul populations was the absence of freedom[111] in every sense and above all, the priority activities; food crops, small livestock, hunting, fishing and traditional crafts were influenced.

This has led to food insecurity in the families. As far as the scope of this resistance is concerned, we can highlight the social organisation that has characterised the Mandoul region, with lineage-based societies grouped together in villages where autonomy reigned. This hinders any form of solidarity between different ethnic groups in a common cause. As a result of the reluctance and resistance shown by the local people, a number of favourable measures were taken in their favour by the colonial administration, including bonuses. Christian Bouquet quotes Chevalier and Senay[112] as saying:

Gradually, the cotton companies made efforts to transform the mentality of the black man. To help them with their work, they distributed hoes; to reduce the need for portage, they increased the number of buying and ginning centres and developed means of transport; they distributed salt and incentives[113]
.

Premiums were awarded to exemplary farmers who looked after their cotton fields well. This was to encourage farmers to take an interest in cotton growing, despite the forms of resistance they put up. The colonial administrators were intent on exploiting the farmers, even though they had granted them aid to develop cotton and food crops. This is why these agents had thought of setting up training for producers.

3. Supporting producers in growing cotton

Faced with the attitude of the local people, who did not welcome the compulsory cultivation of cotton, and the lack of technical support staff, the colonial administration asked the traditional chiefs to contribute. This was achieved through the introduction of incentives paid to village chiefs. The practice of this supervision was characterised by brutality in the form of forced labour, whipping by the "*gou- miers*", i.e. the chief's bodyguards, tarnished "the image of cotton growing in Chad until 1944, the official date for the abolition of forced labour"[114] . According to the official report from COTONTCHAD-SN and the Ministry of Agriculture, the oppressed people were appeased by COTONFRAN agents, the colonial administration and traditional chiefs, and cotton growing was gradually integrated into farmers' production

[110] Interview with Mognara, Ngomanan, 24 June 2017
[111] Interview with Sillenengar, Ngomanan, 24 June 2017
[112] P. Senay, 1949, p. 283.
[113] C. Bouquet, 1982, *Tchad, genèse d'un conflit*, Paris, L'Harmattan, p. 88.
[114] Interview with Tchadingar Halinan, Koumra, 19 June 2017

systems 12 years later after the abolition of forced labour[115] .

Technical research was also useful for scientific seed studies. The first cotton seeds came from Congo-Belgium. The administration set up an AEF cotton committee responsible for introducing and selecting plant material. This committee undertook the first cotton trials in 1931 and created the karoual (Gounou- Gaya), Tikem (Fianga) and research station cotton selection and trial farms in 1936. The Tikem farm was converted into a research station in 1937. Following this, the Institut de Recherche Cotonnière et des Textiles exotiques (IRCT) was set up in 1946 and then moved to Bé- bédjia[116] . The colonial administration had also given thought to intensifying cotton growing for rural development. As part of the Fonds d'Investissement pour le Développement Économique et Social (FIDES), hydraulic projects were implemented to dam the Logone river from Laï to Katoa and to develop irrigated cotton paddocks. However, the first harvests in 1957 were disappointing due to the unsuitability of the soil for this crop[117] . As a result, cotton growing has remained exclusively rain-fed in Chad until now.

B. The development of cotton production during the colonial period

The trend in cotton production varies from crop year to crop year and is known by the tonne. It is due to the commitments of the farmers and the cotton company.

1. Cotton production in tonnes from 1928 to 1959

Production, which was very low at the start of industrial cultivation, grew gradually but slowly. This growth became rapid towards the end of the 1940s when farmers began to embrace cultivation. From 17 tonnes in 1928, production rose to 1,200 tonnes after two years (1930) and to 7,500 tonnes in 1935. Ten (10) years later (1945), it reached 42,700 tonnes and more than 53,000 tonnes in 1949/50 at a yield of 297 kg/ha[118] . This increase in production is essentially due to the extension of land under the influence of forced labour. The statistics given above explain the change in the area under cultivation[119] . The area subsequently maintained an upward trend thanks to the farmers' commitment to this crop. However, production has fluctuated, while maintaining an upward trend, reaching 94,000 tonnes in 1953/54 and peaking at 149,000 tonnes in 1959/60 at a yield of 503 kg/ha. This remarkable increase in yield can be explained to some extent by the productivity and agricultural supervision programme implemented by the ONDR from 1965 onwards[120] .

2. The evolution of seed cotton prices during the colonial period.

Given that the price of cotton on the world market was subject to wide fluctuations as a result of overproduction, strong market demand or international political crises, cotton producers could not tolerate these fluctuations, especially if prices collapsed. So in 1946, the colonial administration set up a fund to stabilise cotton prices. In addition to developing cotton growing in Africa, this fund was intended to regulate cotton prices and guarantee countries a fixed rate despite fluctuations in world prices[121] . Its creation coincided with the cotton boom

[115] Prepared by COTONTCHAD-SN and Ministry of Agriculture, 22 February 2016, p. 5

[116] *Ibid,* pp. 6-9.

[117] Struizinger Ulrich, Tchad, 1983, "Mise en valeur", *Coton et développement, Tiers-Monde*, Année 1983, Volume 24, Numéro 95, p. 87.

[118] F. Nuttens, 2001, La production du coton graine en zone soudanienne, N'Djamena, Ministère de l'Agriculture, ONDR/DSN, p. 32.

[119] *Ibid.*

[120] *Ibid.* p. 34.

[121] For example, when the purchase price from producers is 20 FCFA per kg, the F.O.B. price is 125 FCFA. This is the cost price delivered to the port of Pointe-Noire; the CIF price is 125 CFA francs, which is the price of

on the world market, which was stimulated by the strong demand for cotton at the end of the Second World War and above all by the Korean War[122] . As a result, prices remained firm until 1952. At the same time as this period of prosperity, the purchase price of seed cotton from Chadian producers had risen considerably.

The price-setting mechanism for seed cotton under CO- TONFRAN took account of the price policy, which guaranteed income for producers and encouraged them to continue producing cotton, thereby generating foreign currency for the State. The purchase price was 1 FCFA/kg in 1930 and rose very slowly to 3 FCFA between 1945 and 1946[123] . The price of seed cotton rose from 12 FCFA four (04) years later to 16 FCFA/kg in 1950/51. It then rose to 24.20 FCFA/kg in 1956/57, at which level it was artificially maintained for 14 marketing years to compensate for input subsidies, before rising again to 26 FCFA from 1958-1959[124] . This mechanism for setting seed cotton prices worked with the main players in the cotton sector, according to the standards set by them. Prices for seed cotton changed gradually from 1931 to 1948. The table below shows changes in seed cotton prices on self-managed markets throughout Chad.

Table 1. Trends in seed cotton prices from 1930 to 1959

Years	**Price/Kg**	**Years**	**Price/Kg**
1930	**1**	**1945**	**3**
1931	**0,7**	**1946**	**4**
1932	**0,7**	**1947**	**5**
1933	**0,6**	**1948**	**12**
1934	**0,6**	**1949**	**12**
1935	**0,6**	**1950**	**16**
1936	**0,75**	**1951**	**25**
1937	**0,85**	**1952**	**25**
1938	**1,1**	**1953**	**24**
1939	**1,1**	**1954**	**24**
1940	**1,1**	**1955**	**24**
1941	**1,1**	**1956**	**26**
1942	**1,5**	**1957**	**26**
1943	**2,25**	**1958**	**26**
1944	**2,5**	**1959**	**26**

Source: F. Nuttens, (2001)

From 1953 onwards, the steady upward trend in producer purchase prices that had been underway since the end of the Second World War was interrupted by the collapse of world cotton prices as a result of inflation in mainland France in 1951, which slowed the growth of cotton sales. In addition, the fall in international prices led to the 1952 crisis and the fall in

cotton fibre delivered to Le Havre; the selling price of 300 francs is the selling price on the world market. According to A. KOTOKO, the support fund receives 80% of the difference between the selling price and the CIF price, i.e. 140 CFA francs per kg. When there is a fall on the world market, this sum is used to top up the purchase price paid to producers.

[122] W. Stevelinck, 1953, "Le développement du coton dans la zone Mayo-Kébbi, Logone et Moyen Chari, marchés coloniaux du monde", n° 399, C.A.O.M, p. 408.

[123] F. Nuttens, 2001, p. 62.

[124] *Ibid.*

wholesale prices[125] , which came to an end in 1953 with the outbreak of the Korean War. Since then, prices paid to producers have depended on the quality of the cotton. 1er grade cotton was worth 25 CFA francs per kg, while 2nd grade cotton cost 20 CFA francs per kg[126] . This new pricing system applied in Chad during this period was determined by the selling price on the Havre markets. In fact, when the *Allen* cotton grown in Chad was of first quality, it benefited from a premium of 200 points on the Havre markets, i.e. around 7.60 CFA francs per kilo. However, the proportion of second quality seems to have been the highest in Chadian production. Nevertheless, cotton purchase prices fell by an average of 2% between 1953 and 1960 compared with 1952, due to the classification of prices according to cotton quality. As a result, spinning mills placed more emphasis on consistent quality and homogeneous batches than on paper, the second or third choice, their problem being to adjust their machines and determine their production according to quality[127] .

However, assistance from mainland France took the form of increased and continuous intervention, in the form of the FIDES. In 1954, the Caisse de stabilisation des prix de l'AEF replaced the Caisse de soutien and, in November 1956, the Fonds de soutien des textiles d'Outre-Mer was created, funded by a 30% rebate on the textile tax (itself funded by a 0.70% tax on all imported cottons)[128] . The fund then focused on price stability at producer level, possibly taking a portion of the profits from cotton sales by cotton companies when prices allowed[129] .

The purchase price to producers was stabilised at between 23 and 25 CFA francs per kg, despite the fall in world prices. This is because a drop in price would not have been understood by producers and would have reinforced farmers' neglect of cotton growing:

We need to take into account the particular psychological reactions triggered by the announcement of a drop in the purchase price of their products in the case of uneducated producers, who are more sensitive to the contribution of integrity between themselves and European buyers. They immediately believe themselves to be the victims of a manoeuvre and behave accordingly[130] .

It was therefore necessary to stabilise purchase prices to avoid a decline in cotton cultivation among growers who, moreover, were constantly questioning its consequences. However, if the result of sales, instead of showing a profit, showed a deficit, this was shared equally between the fund and the company. This mechanism enabled the Caisse de Stabilisation to build up reserves in anticipation of bad years and limit the company's losses when prices fell sharply.

3. The production profit-sharing policy in the face of the cotton crisis

The Caisse de stabilisation acted as a price regulator, ensuring the annual payment of a sowing premium and the implementation of production credits designed to improve yields per hectare. Sowing premiums were paid at the rate of 900 CFA francs per hectare, subject to a deadline set each year by the government[131] . The aim of this measure was to encourage early sowing, which is one of the conditions for good plantation yields.

[125] F. Chapronnier, 1959, *La crise de l'industrie cotonnière française*, Paris, Génin, p. 359.

[126] G. Sautter, p.128.

[127] W. Stevelinck, op cit, pp.1947-1949, C.A.O.M., p. 408.

[128] Secretary of State for Community Relations, October 1960, p. 12.

[129] *Ibid.*

[130] S. Gilles, p. 129.

[131] G. Digambaye and R. Langue, 1969, *L'essor du Tchad*, Paris, p.132.

C. The organisation and operation of cotton production

1. Organising farmers

In the cotton-growing area of Chad, there were traditionally informal groups made up of members of the same family or people who were grouped together by affinity for the purpose of mutual aid or working together. In 1984, with the change from an individual to a group approach, and in application of the measures set out in the contract, the ONDR began organising producers into "input groups", which later became "producer groups"[132] . In 1986, given the plethora of producer groups to follow, the ONDR decided to set up Village Associations (AV) on the basis of the producer groups. Coinciding with the end of input subsidies from the European Union, it imposed the setting up of cotton growers' associations in the villages, and by the end of 1992, all the villages in the cotton-growing zone were organised[133] . The aim of this structuring was to transfer responsibilities to producers through their associations. As a result, the systems for managing inputs and marketing seedcotton changed. Producers became stakeholders through their associations, to which the following activities were transferred:

> Census of agricultural input requirements ;

> Receipt of inputs and distribution to members ;

> Marketing of seed cotton through self-managed markets (MAG) ;

> Collection of income from cotton and payment to producers.

However, this organisation expanded under the impetus of the technical services of the Ministry of Agriculture, and in 1992 the producers were reorganised into the Mouvement Paysan de la Zone Soudanienne (MPZS). The MPZS wanted to be a federation of farmers' associations at national level and a partner of the cotton company. It is remunerated on the basis of 300 FCFA/tonne of cotton marketed, deducted from the fees paid to the co-tillers 3,500 FCFA/tonne of cotton marketed in return for the activities undertaken[134] . But the MPZS is more like a trade union. It does not report to its members on its activities or the use of its resources. In addition, the membership renewal elections provided for in the articles of association never took place, and the movement was sometimes dominated by traditional chieftains[135] .

In 2000, as part of the implementation of measures to accompany reforms in the cotton sector, a new restructuring led to the creation of local coordination committees (CCL)[136] . The restructuring began with awareness-raising campaigns among producers, followed by democratic elections for village delegates and then cantonal delegates (December 1999-March 2000). Based on the cantonal delegations (DC) as members by right, Local Coordination Committees (CCL) were set up by election around each ginning factory zone from 25 April to 05 May 2000[137] . However, the ten (10) LCCs joined forces to form a national umbrella organisation called the "Union Nationale des Producteurs de Coton du Tchad (UNPCT)" on 02 April 2007. The UNPCT takes part, along with the other players in the sector, in meetings to determine the parameters of the cotton season (setting the price of seed cotton and inputs), to examine offers of inputs and to prepare the marketing of seed cotton. It participates in the

[132] Interview with Rimtoibaye, Koumra, 06 June 2017

[133] Interview with Rimtoibaye, Koumra, 06 June 2017

[134] COTONTCHAD activity report, p. 23.

[135] *Ibid.*

[136] CTRC, 2006, Rapport de mise en place des comités de coordination locaux (CCL), p. 46.

[137] *Ibid.*

revision of the MAG charter and in all activities involving the participation of players in the sector.

COTONTCHAD also takes input orders through its "field agents, who act as an "interface". These orders are based on the land to be cultivated, as estimated by the farmers"[138] . The cost of inputs is then deducted from income. This system has been replaced by the work of the supervisory agents and the ONDR. However, its operation leaves much to be desired insofar as input markets are practically non-existent and COTONT- CHAD cannot meet all farmers' needs. This rationing of inputs leads to certain practices whereby farmers "dilute" inputs by using them on other crops, or by applying them to larger areas than those recommended for their effective use.

2. The functioning of the cotton sector in Chad

The operation of the cotton sector takes into account the management of skilled labour and the role played by the players. The 1949 conventions required cotton companies to modernise their factories[139] . Until then, most operations were carried out manually: feeding the gins, collecting the cotton fibre from the gins, pressing 10 kg out of 10 kg of the fibre to be pressed, grooming the bale before pressing, wrapping and strapping the bales, rolling the barrels of water to feed the locomobile[140] .

However, thanks to pneumatic suction and the creation of pump-fed water towers, the supplies introduced between 1952 and 1955 made it possible to reduce handling operations when unloading seed cotton lorries, wrapping and strapping bales and storing them. As a result, the number of seasonal labourers hired by COTONFRAN for the ginning period (December to May) was reduced by extending its sector from the south, i.e. the Sara country[141] .

The new installations required staff who had already been trained, and it was only natural that the first installations of what has sometimes been referred to as the "colonisation" of Chad by the Sara people should focus on them[142] . The permanent employees are housed by the company. Seasonal workers were recruited locally from the local farmers, who were brought together by invitation to tender. Very often, when the population was limited, they were hired by the canton chiefs that the company was interested in. These labourers did not like to remain employed beyond the agreed time. The wage for a labourer is 65FCFA per working day. Permanent employees received salaries of around 2,000FCFA for the planter and 15,000FCFA for the mechanic[143] . To better manage the Chadian cotton industry, COTONFRAN has set up ginning plants in the various regions of southern Chad.

III - THE CREATION OF THE GINNING PLANT IN KOUMRA

The Koumra ginning plant is located in the capital of the Mandoul Oriental region in southern Chad. It was created in 1934 and was initially managed by COTONFRAN, a company founded in 1911. The Koumra factory zone is located between the Sarh factory to the east and the Doba factory to the west. It is bordered to the north by the Tandjilé East region and to the

138
139Interview with Beramgoto, Koumra, 08 June 2017 G. Magrin, 2001, p. 53.
Ibid.

140
141 J. Chapelle, 1986, *Le peuple Tchadien, ses racines et sa vie quotidienne*, Paris, L'Harmattan, p. 104.
142 J-P. Magnant, 1987, *La terre Sara, terre tchadienne*, Paris, L'Harmattan, p. 65.
143 COTONTCHAD, 2009, Rapport des campagnes agricoles et commerciales, pp. 7-8.

south by the Central African Republic (CAR)[144] . The northern part of the Kou- mra plantation has wooded savannah vegetation dominated by the shea tree, which is highly protected by the local population. The south is characterised by wooded savannah, plains and a large expanse of forest that extends as far as the Central African Republic, making it ideal for growing cotton and rearing cattle. Much of the southern zone is made up of valleys and flood plains that are ideal for growing cash crops, small millet, sorghum, groundnuts and rice. Agriculture and livestock farming are the two main activities in the region. The soil is less clayey and much of it is ferruginous, which degrades easily if cultivation practices are not properly applied. The Koumra ginning plant, formerly managed by COTONFRAN, came under the management of COTONTCHAD in 1971 under a memorandum of understanding between the Republic of Chad and COTONFRAN. COTONTCHAD is a semi-public company under Chadian law. Its turnover[145] was six hundred million CFA francs, which has now risen to five hundred billion ten million CFA francs. COTONTCHAD has its management based in Moun- dou and its head office in N'Djamena. The company provides a livelihood for more than three million Chadians, including cotton growers, insurance companies, banks, transport companies and traders. Every year, it distributes more than thirty billion CFA francs in revenue, making an effective contribution to the national economy and, approximately, more than thirty-five billion CFA francs in turnover, which is not negligible as the country's currency. It pays a substantial sum to the State in various forms: taxes paid to banks in the form of interest and financial charges[146] . Unfortunately, the company has been going through an unprecedented financial crisis in recent years as a result of the crisis experienced by the African cotton subsidiary on the world market. However, following the presentation of the Koumra factory and COTONTCHAD, our study focuses on the privatisation process of the Chad cotton company.

A. The process of privatising the cotton company

The establishment of the Franco-Belgian cotton company COTONFRAN in 1934 was an opportunity for the company to set up ginning factories in most of the so-called southern regions, in the major towns of Chad. At the time, cotton was considered a major economic and commercial issue at national level.

More than forty years later, the colonial cotton company CO- TONFRAN was to cease operations in Chad. A memorandum of understanding was signed between the Republic of Chad and its partners, and COTONFRAN was made possible. The signing of this memorandum of understanding gave rise on 29 October 1971 to the Chad cotton company known as "COTONTCHAD"[147] . A few decades after this agreement was signed, this company was shaken by the international crisis in the cotton sector and the mismanagement that arose after the 2000/2001 campaign, the capital of which amounted to 6.6 billion CFA francs. The company continued to operate in breach of Articles 664, 665 and 667 of the Uniform Act on Commercial Companies in the OHADA region, reaching negative equity of more than 48 billion CFA francs at 31 December 2010.

---

[144] E. Mbainaissem, 2013, "Audit stratégique de la cotontchad-sn" Dissertation, Centre Africain d'études Supérieures en Gestion, p.105

[145] Wawe Harouna, 2015, "Pratique de contrôle budgétaire, COTONTCHAD-SN" Professional Master's thesis, University of Ngaoudéré, p. 47.

[146] Ibid.

[147] Rapport (Tchad), 1971, Protocole D'accord entre la COTONTCHAD, ONDR, IRCT, DGRHA, DRHFRP et FIR, p. 7.

However, the process of restructuring COTONTCHAD was triggered by the restructuring plan presented and approved in October by Decree No. 024/PR/PM/SGC/DG/PMI/2011 of the Council of Ministers of Thursday 29 September 2011. Subsequently, letter No. 391/PR/PM/MCI/SG/DIAPME/PMI/2011 dated 05 October 2011[148] , from the Minister of Trade and Industry in his capacity as supervisor and representative of the majority shareholder of COTONTCHAD, requesting the Chairman and CEO of this company to take all necessary steps to implement the restructuring process[149] .

The minority shareholders (GEOCOTON, SGT and ECOBANK) agreed to sell their shares for a symbolic franc to the majority shareholder (the State), triggering the liquidation process of COTONTCHAD with the appointment of a liquidator. This process was approved by the Extraordinary General Meeting of shareholders on 30 December 2011[150] . Following the reading of the report by the Board of Directors, the limited company known as Société Cotonnière du Tchad, Société Nouvelle[151] , abbreviated to "COTONTCHAD-SN", was created and all the following formalities were completed:

> Setting the share capital at 5010000000 (fifty billion ten million) FCFA divided into 501,000 (five hundred and one thousand) FCFA shares with a par value[152] of ten thousand (10,000) FCFA each, paid up at the time of subscription;

> The adoption of COTONTCHAD-SN's articles of association on 12 April 2012;

> The appointment of a Managing Director and a Deputy Managing Director.

1. The objectives of COTONTCHAD-SN

- The purchase of seed cotton, its ginning and the marketing of cotton fibre and cotton by-products;
- Participation in all agricultural development operations in Chad, both industrial and commercial, and in terms of production;
- To be directly or indirectly involved in all transactions whose corporate purpose is likely to facilitate the achievement of such purpose.

2. The mission of COTONTCHAD-SN

The mission of this new company is :

> Exclusive marketing of cotton (purchase and transport of seed cotton, ginning, packaging, sale of cotton fibre and cotton by-products) produced in Chad;

> To participate in the development of agricultural production and productivity in liaison with organisations, production charges and rural promotion;

> To undertake to buy all the seed cotton brought to the market by individual growers or delivered by grower groups.

B. The organisation chart of the Koumra ginning plant

The Koumra plant is organised as follows:

- A Cotton Zone Co-ordinator (COOZOC)[153] , who reports to the CEO, co-ordinates and

---

[148] Société cotonnière du Tchad, 2011, Présentation du plan de restructuration COTONTCHAD Société Nouvelle, p. 11.

[149] *Ibid.*

[150] Document de stratégie de reformes du secteur coton tchadien, 1999: adopte par le Haut Comité International. République du Tchad Comité Technique du HCI, p. 52.

[151] Réforme de la filière coton au Tchad, 2004: Etat d'avancement. Cellule Technique Reformes Coton (CTRC), p. 15.

[152] P. Honorat (2009), *le budget facile pour les managers:* démarches, indicateurs, tableau de bord, Ey- rolles, Paris, Paperback, p. 230.

[153] A. K. Mahamat, 2014-2015, Rapport de stage à la cotontchad-sn, usine de koumra (Tchad), p. 8.

administers the factory's activities in general and solves problems within the company. He has under his control a management controller, an office secretary and a radio operator;

- A Plant Manager (PM), who is assisted by a Team Manager, Maintenance and Safety (TMMS) and has the following shift supervisors reporting to him;
- A supervisor in charge of production management, with a limited role between the factory and the cotton growers. He is assisted by two trainers, fourteen Cotton Field Agents (CFAs) and an input storeman. The supervisor's role is to mobilise the ACTs to collect statistical data from the cotton fields, assess production in relation to productivity, draw up a purchasing schedule for each village association (VA), summarise the data collected and centralise it in order to draw up a plan for the introduction of production inputs;
- A Weighbridge Manager (GPB) who in principle reports to the Cotton Zone Coordinator (COOZOC). He is supported by a fleet team leader, a mechanic who takes part in activities during the campaign and a shift weigher who frequently assists with activities[154] . His main activities are :
- Rational management of weighbridge activity;
- Ensure the smooth running of logistical equipment;
- Unload the tickets at the COOZOC after weighing the vehicle;
- Find out about all transport equipment and fuel consumption per vehicle.
- An Administrative and Accounting Manager (RAC) is assisted by a cashier and parts storekeeper. The RAC pays the wages of workers and seed cotton growers.

The Koumra plant employs more than "two hundred and forty-four (244) employees, including two (02) managers, fifteen (15) supervisors, eighty-one (81) permanent workers and one hundred and forty-six (146) seasonal workers"[155] . Its main activity is to:

- Bringing the seed cotton from the buying centre to the factory;
- Pay the money for cottonseed to the VA;
- Renew input and treatment equipment credits;
- Ginning seed cotton and packaging bales of cotton fibre;
- Receive quality inputs and distribute them to VAs ;
- Shipping finished products (cotton fibre) from Ngaoundéré to the port of Douala (Cameroon) to be sold on the world market;
- Shell the cotton seeds and send the fines to the Moundou soap mill.

In addition, the Koumra factory is supplied by part of the Doba and Sarh cotton-growing zones, due to its seed and commercial activities. It has fourteen (14) input management centres covering twenty-nine (29) cantons and more than sixty Village Associations (VAs) in the Mandoul Oriental region[156] . The structure of the COTONTCHAD-SN ginning plant in Koumra is described below.

Figure 2. Koumra ginning plant management chart

[154] Interview with Pafing, Koumra, 16 June 2017

[155] Interview with Pafing, Koumra, 16 June 2017

[156] COTONTCHAD-SN, 2009, Rapport d'activités agricoles et commerciales, campagne 2008/2009.

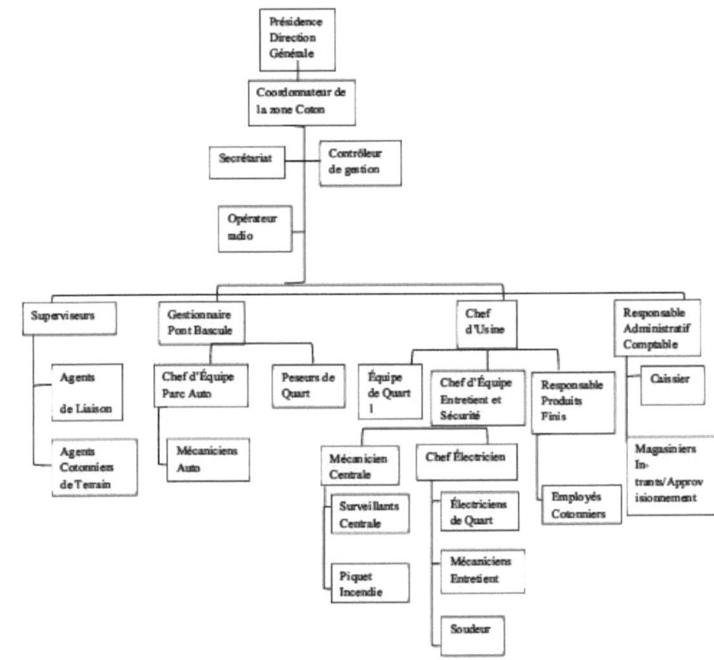

Source: Report, COTONTCHAD-SN (2017)

C. Plant operations in the various departments

The Koumra ginning plant has been managed by COTONTCHAD since 1971 as a mixed company under Chadian law, known as COTONT- CHAD-SN on 11 November 2011. It is worth analysing its operation through the various departments within it.

1. Administration and maintenance department

The administrative department includes a large number of people responsible for maintaining commercial relations with head office and other departments. This department is also responsible for finding potential new customers for the company and managing staff[157] . The maintenance department employs a team of people who are generally divided into different sub-departments. The maintenance team is on call 24 hours a day, 7 days a week to ensure that the plant's operations run smoothly. To do this, the technicians rotate night and day. There is a mechanical workshop, with three people in charge of monitoring the machines used for ginning seed cotton and cotton fibre. In the electrical workshop, there are four people responsible for managing all the company's electrical installations, including wiring the robotic systems[158] .

2. General shop department

The general shop controls the stock shop, the parts shop and the input shop. As far as the stock warehouse is concerned, we studied the premises that enabled the Koumra ginning plant

[157] E, Mbainaissem, 2013, p.76.
[158] *Ibid.*

to store seed cotton and cotton fibre in the event of saturation. Secondly, the parts warehouse stores all the plant's equipment, which its employees use when they need it. Finally, the input shop is used to "store inputs at reception before distributing them to the VAs"[159] . The input storekeeper has to manage cotton production supplies with great care, noting down the CO-TONTCHAD order slip on receipt. He is responsible for taking the inputs out of the shop, noting down the dispatch note, which bears :

The vehicle number, product code, destination, VA code, product unit, product quantity, unit price, valuation and reserve. The dispatch note also bears the name of the driver and the name of the forwarding warehouseman, the date, and the signature of the chairman of the VA and the driver for confirmation. There is also the receiving slip, which shows the vehicle number, the name of the driver, the carrier, the order number, the supplier, the delivery slip number, the date, the movement slip number with the date and the product code[160] .

In this slip, we mentioned the part reserved for the activity manager for approval, his surname and first name, the date and signature. There is also the "section reserved for the shop clerk to take charge of the goods after checking, which includes his first and last name, date and signature"[161] . Therefore, following the general shop service, we have identified the packing service and the parking service. The packing service is semi-automated, with labourers always on standby. The packing speed is superb, as is the organisation. In this department, there are two (02) technicians in charge of weighing the bales after they leave the machine. They note the net weight of the bales and take samples of the cotton fibre.

The parking service is provided by the Weighbridge Manager (GPB)[162] . The BPG keeps an eye on lorries full of cottonseed on the weighbridge in order to ensure that the weight is correct. He recommends that drivers "place the empty lorry back on the weighbridge after unloading the cottonseed and weigh it again to obtain the net weight of the cottonseed"[163] . Since then, all the departments have worked to ensure that COTONTCHAD runs smoothly by sharing tasks and specialising their work.

In short, the Chadian territory has known different crops that have promoted economic and social development before the compulsory cultivation of cotton. The colonial administrator used this crop as an economic and political granary to dominate the local populations. Ignorance of the system and methods of cultivation was a new turning point for the local people, who were to succeed by force for the benefit of the colonial administrator and the traditional chiefs. In truth, the introduction of cotton cultivation changed the rural world through the infrastructure and agricultural equipment loans granted by COTONTCHAD and the Office National pour le Développement Rural (ONDR) to producers, even though the purchase price of seed cotton was falling. This weakened the country's economy as a whole, following crises in world prices, and led to the privatisation of the Chadian cotton industry.

[159] Interview with Didjenbaye, Koumra, 15 June 2017
[160] Interview with Didjenbaye, Koumra, 15 June 2017
[161] Interview with Didjenbaye, Koumra, 15 June 2017
[162] C. Nantiga, 2013, Rapport de stage, usine de Koumra (Tchad), p. 9.
[163] Interview with Oumar Hissein, Koumra, 15 June 2017

CHAPTER II

COTTON PRODUCTION IN THE EASTERN MANDUL REGION

The history of cotton in Chad is well known, as a forced crop of colonial origin. Cotton has played a major role in the national economy. It remains the country's only commercial speciality, in that production is entirely destined for export. From its origins to the present day, cotton production has been characterised by an upward trend throughout its history[164] . Since the 1980s, these irregularities have tended to become more pronounced as a result of serious cotton crises (1984-1999). These were mainly the result of falling world prices for cotton fibre, poor management of the sector and chronic insecurity perpetuated by politico-military unrest[165] . The abolition of subsidies on inputs and agricultural equipment has had a profound effect on Chad's cotton sector, and is one of the main measures taken. The main aim of these measures is to turn the sector around, which means, first and foremost, reducing the role of the state and the supervisory authorities in cotton production.

In fact, the supervision of farmers, initially the responsibility of the ONDR, has now been made more flexible. The creation of the interface and the structuring of the farming world are intended to take over this role from the ONDR. Village associations, set up on the initiative of the cotton company with the support of the ONDR, now manage inputs and organise the primary marketing of seedcotton on self-managed markets. While this process, which is moving towards privatisation of the sector, seems to be working well in some cotton sectors, this is not the case in others[166] . Producers' poor organisational capacity and illiteracy are sometimes a source of difficulties in taking charge of the production and primary marketing of cotton. To do this, it is important to present the factors and players involved in seed cotton production.

I - PRODUCTION FACTORS AND PLAYERS SEED COTTON IN EASTERN MANDOUL

A- Production factors for seed cotton

Thanks to its geographical location, the Mandoul Oriental region is rich in natural resources that are favourable to cotton growing. The production of seed cotton is based on natural, human and chemical factors.

1) Natural factors

Natural factors are those elements of nature that favour the cultivation of cash crops and food crops. Among these factors, we studied the soil, climate and vegetation.

a) Soils

The Mandoul Oriental region has two types of soil: tropical ferruginous soil and ferralitic soil, suitable for growing cotton, groundnuts and millet[167] . Initial soil studies carried out in collaboration with ORSTOM analysed the different soils[168] . They highlighted the differences in their mineral content in relation to the specific needs of the cotton plant.

G. Magrin, 200, p. 52.

Ibid.

Ibid. p.60.

[167] G.T.Z, 1988, *Analyse régionale sommaire du Mandoul Oriental (Tchad)*, p. 5.

[168] *Ibid.*

b) Climate and vegetation

Due to its geographical location in the south of the country, the Mandoul region belongs to the Sudano-Guinean climate zone. This area is characterised by fairly abundant rainfall[169] . The isohyets range from 800 millimetres to more than 1,000 millimetres in a normal rainfall year, enabling good cotton yields[170] . As in the rest of the country, the Mandoul Oriental region has two seasons: a rainy season and a dry season. The rainy season begins in May and ends in October. The dry season runs from November to April[171] . This situation determines the activities of the local population, which are almost entirely dependent on the rainfall recorded. It is worth mentioning that agriculture is the main activity practised in the region, and people base their agricultural calendar on this natural mechanism[172] .

Temperatures also vary considerably depending on the time of year. They reach their highest level in March, peaking at 37°C, and drop to around 17°C in December. Temperatures rise as you move from the south of the country towards the north, where the maximum temperature easily reaches 45°C in the shade[173] .

In terms of vegetation, the area is densely forested with a multitude of tree species (shea, néré, acacia, jujube and caïcé- drat). This forest is characterised by a wooded savannah and, in some places, a more or less dense forest[174] .

2) The human factor

The human factor has played a significant role in agricultural activities for local populations. Man, as a creative genius, is a means of all kinds, hence the skilled labour needed to grow cotton. In Man- doul, as elsewhere, farming activities are carried out as a "family", starting with clearing the land to be cultivated, sowing, weeding, tending, harvesting and sorting. It should be noted that women and children make up a large workforce compared to farmers"[175] . In some cases, the farmers sought to employ the workers in compensation, either with money or with gifts in kind[176] .

3) The chemical factor

The chemical factor in cotton growing comprises all the products needed for the intensive cultivation of cotton and its by-products. These include the spreading of NPK fertiliser and the treatment of cotton plants with phytosanitary products, an operation specific to cotton growers[177] . It should be remembered that the acquisition of these products as part of an integrated production chain is conditional on cotton production. Only cotton growers use the specific chemical products distributed by COTONTCHAD. Cotton growing requires fertilisers, herbicides and treatments[178] .

Pesticides and insecticides are essential to combat the pests found on the plantation, and farmers are often powerless to do anything about them because other control techniques are of limited effectiveness. To combat pest attacks or the proliferation of diseases, it is advisable to

---

169 J. Cabot and C. Bouquet, 1974, *Géographie, le Tchad*, Paris, Hatier, p. 96.
170 *Ibid.*
171 *Ibid.*
172 F. Hautcoeur, 2001, *Guide pour l'accompagnement des ruraux dans la gestion conservatoire de leur espace*. G.T.Z, p. 12.
173 *Ibid.*
174 G.T.Z, 1988, p. 9.
175 Interview with Djainabaye, Kemkian, 09 June 2017
176 Interview with Alladoumngue, Kol, 11 July 2017
177 Kibassim Bagrim, 1975, "Agriculture commerciale, modernisation et développement rural en zone cotonnière, exemple du Moyen-Chari", PhD thesis in Development Sociology, University of Paris V, p. 86.
178 *Ibid.*

practise crop rotation, alternate crop rows or use crop combinations. Because very few pests or diseases affect young plants[179] . The use of chemical products is a sign of modernism and crop intensification in Chad, particularly in the Mandoul Oriental region. In addition, cotton growers bring manure to maize fields in rotation with NPK-fertilised cotton fields. This maintains soil fertility for the cotton on which it is spread, and also for the cereals planted the following year[180] . In this way, seed cotton production takes account of the stakeholders involved in its organisation.

B- Players in seed cotton production

The players involved in seed cotton production in Chad and the Man- doul Oriental region are as follows:

- Cotton producers: The primary role of producers is the individual production of seed cotton and the delivery of their production to COTONT- CHAD-SN through their organisations known as Association Villageoises (AV) in accordance with the clauses of the agreement known as the Charte de Marché Autogéré (MAG). The VAs are the direct respondents of COTONTCHAD-SN in activities such as the provision of inputs on credit to producers, the marketing of seed cotton and debt recovery. They are set up as Local Coordination Committees (LCCs) on the basis of cantonal delegations, which have formed a national umbrella organisation, the Union Nationale des Producteurs de Coton du Tchad (UNPCT). The UNPCT and the CO- TONTCHAD-SN[181] sit down with the government to set prices for seed cotton and inputs, and to examine offers for inputs. The UNPCT is also involved in marketing seed cotton, revising the MAG charter and all activities involving producer participation. Overall, relations are good between the two partners[182] .
- ONDR: responsible for providing support to the rural world as a whole
for all agricultural crops, including cotton. It is a member of the commission responsible for resolving problems relating to the quality of seed cotton in ginning plants. The partnership between ONDR and COTONTCHAD has been established through a collaboration agreement and a memorandum of understanding, the latter of which were signed by the two parties[183] on 03 and 13 May 2015 respectively.
- ITRAD: a public scientific and technical establishment
responsible for research into plant production, including cotton (varietal research, basic seed production, conservation of phylogenetic resources and development of crop technology packages).

The partnership between ITRAD and COTONTCHAD is based on an agreement that formalises the framework for the provision of services and expertise to COTONTCHAD. These services include contributing to the implementation of the seed plan, the production of pre-basic and basic seed with a view to maintaining varietal purity, the training of COTONTCHAD field agents, seed multiplication farm staff and seed producers on topics relating to cotton seed production[184] . They also include drawing up scientific and technical

---

[179] O. Mandi, 2007, "Coton culture et mutation socioéconomique dans la zone soudanienne au Tchad de 1928-1999", Master's thesis in Economic History, University of Yaoundé I, p. 74.

[180] *Ibid.*

[181] CTRC, 2006, Réforme de la filière coton au Tchad. Etat d'avancement. Rapport d'activités, p. 24.

[182] Macra Tadin, 1983, "L'intervention de l'État dans le secteur cotonnier au Tchad", Doctoral thesis in Public Law, University of Toulouse, p. 65.

[183] *Ibid.*

[184] ITRAD, 2007, Rapport d'activités de la production semencière du coton (Report on cotton seed production activities)

recommendations and taking part in meetings relating to calls for tender for agricultural inputs, and carrying out any other expertise relating to agricultural inputs, seed cotton ginning and fibre quality[185] .

- COTONTCHAD-SN: it performs the following functions:
- The supply of agricultural inputs (fertilisers, plant protection products, seeds) to cotton growers for cotton production, mainly on credit;
- The purchase of seed cotton with an obligation to buy the entire production;
- The transport of seed cotton, part of which is subcontracted to private transporters;
- Ginning of seed cotton, evacuation and marketing of cotton lint.

The transport and processing of the seed into oil and cake and their marketing. COTONTCHAD-SN acted as a relay for the ONDR in supervising cotton growers. It was also involved in financing the agreements signed between the other players and itself.

- The State: approves bank loans on behalf of COTONTCHAD-SN,

subsidises the factors of production for the benefit of cotton growers and COTONTCHAD-SN's operations and exempts the cotton company from certain taxes such as Value Added Tax (VAT)[186] .

The banks: they grant COTONTCHAD-SN, under the guarantee of the State, campaign credits for the collection and purchase of seed cotton, its ginning and the marketing of the fibre, and productivity credits for the acquisition of inputs and their distribution to producers. It executes the credit guarantee in the event of the cotton company's inability to repay.

C- The organisation of cotton growing in the Mandoul Oriental

Cotton is a colonial crop that was introduced to Chad a long time ago and requires organisation at both farmer and cotton company level. It is on the basis of this organisation that the players benefit from the good yields that promote socio-economic and political development.

1) Organising farmers

The organisation of farmers in cotton growing took place with the introduction of the self-managed system, whereby the cotton companies left cotton production to the farmers. In each village, the farmers meet in groups, each of which has a representative. Above the groups, there is a president known as the "President of the Village Associations"[187] . In fact, before the start of the agricultural season, the president of the Village Associations has to draw up a list for each group in the village, where the group representatives are obliged to record the first and last names of each member, the order and the name of the group. After that, the group representative had to calculate the number of fertiliser orders and members before handing the document over to the VA chairman. The latter, in turn, must calculate the number of all the groups, including their fertiliser orders[188] .

However, this preliminary work is generally carried out in August and September, and the president of the Village Associations will try to contact the cotton agent in the field for the summary and submit it to the COTONTCHAD supervisor[189] . The number of fertilisers

in Chad, p. 15.

[185] G. Magrin, 2001, p.53.

[186] E. Mbainaissem, 2013, p.84.

[187] Interview with Deouyo, koumra, 07 June 2017

[188] C. Arditi, 2004, "Des paysans plus professionnels que les développements? L'exemple du coton au Tchad (1920-2002)". *Tiers monde,* Year 2004, Volume 31, Number 180, p. 68.

[189] Interview with Deouyo, koumra, 07 June 2017

ordered is automatically followed by the number of seeds, phytosanitary products, batteries and equipment for the groups in each Village Association. It is therefore on the basis of the needs of the region's producers that COTONTCHAD will be placing its orders with suppliers of inputs and other agricultural equipment. In addition, the cotton growers will be thinking about field studies, i.e. fertile soils, and cleaning them up to await their orders and the time for sowing. Once the orders have arrived, the presidents of the Village Associations take charge of distributing them to each group. They are assisted by the group representatives, taking into account the list of members and their needs.

2) The organisation of COTONTCHAD-SN

COTONTCHAD-SN has organised itself to make agricultural equipment available to cotton growers. It should be noted that it has placed much more emphasis on the study of seeds in relation to the quality of inputs used during the agricultural campaign. COTONTCHAD-SN, the ginning plant in Koumra, has joined forces with IRCT to produce seeds for farmers. There are several qualities of seed used in the Mandoul region: STAMF Z3T and STAMF Z2T[190] . Initially, Z00 was produced in Bébidjia, Z0 in Békamba, Z2 in Kyabé and Z3 in Koumra[191] . The Z0 produced at Bekamba is combined with the Z1 produced at the Kourmra factory for seed in the zone.

Photo 1: Mixing fungicides with bare hands

Source: Didrot Nguepjouo, 2011, Reproduction: Beyenan Ngarasndi

This photo shows cotton seed mixed with the pesticide before sowing against bacteria.

In addition, there were the STAMF and IRMAB12 varieties popularised by CO-TONTCHAD. The latter has an input storekeeper and a supervisor assisted by cotton agents in the field for monitoring throughout the crop year. It works with the ONDR (Office Nationale de Développement Rural - National Office for Rural Development) to supply certain agricultural equipment to farmers, such as tractors, ploughs (cattle traction) and carts

[190] COTONTCHAD-SN, 2015-2016, Production activity report, pp. 7-9.

[191] ITRAD Bekamba farm, 2014-2015, Seed plan report, p. 13.

27
COTONTCHAD-SN, 2015-2016, Production activity report, pp. 7-9.

28
ITRAD Bekamba farm, 2014-2015, Seed plan report, p. 13.

on agricultural credit. To do this, COTONTCHAD-SN has to draw up the invoice to the presidents of the Village Associations so that, after having made the "seed cotton market" and at the time of payment, it automatically withdraws its money. This is called debt recovery"[29] . Invoices must be handled with great care by the input storekeeper and the COTON TCHAD-SN supervisor.

3) The cotton production system

The popularisation of cotton-growing techniques has led farmers to adopt production systems that are very different from the traditional ones. Scientific analysis shows that intensification has been achieved through the use of improved seeds and mineral fertiliser, and the incorporation of animal-drawn cultivation to save labour at key periods in the cropping calendar and, of course, to facilitate transport. These include crops, crop residues and manure, control of treatments to combat plant diseases and animal manure[30] . In fact, cotton production systems need to be adapted to the production zone, depending on the amount of rainfall during the crop year[31] . Cotton needs 800 to 1000 millimetres of rain for normal growth until it bursts[32] . Again, you have to do the semi within the required timeframe. We have mentioned three phases of sowing:

- Semi-early is the period from 15 May to 10 June;
- Semi normal runs from 11 to 25 June;
- Semi-late starts from 26 June to 10 July.

In some cases, however, cotton has become the dominant, determining crop for which "rotations and techniques are important in areas with a single rainy season. It needs three full months of water for growth and good yields"[192] . Cotton growing remains marginal for other producers, who grow it according to circumstances, particularly in areas with two rainy seasons[193] . In theory, cotton is accessible to all, but it is also a factor of social differentiation. The importance of this crop means that labour is available at various stages: ploughing and sowing, weeding, processing and harvesting.

Generally speaking, the cotton production system takes into account all aspects, including the quantity of seed, which has a greater impact on seedling growth. In fact, two bags of seed must be sown per hectare. Producers are advised to sow cotton in rows with an eighty centimetre spacing between rows and stacks of twenty-five on rich soil and thirty on less rich soil[194] . Weeding and ridging are important for maintaining the cotton plant. The following operations were carried out after emergence:

- The first weeding is accompanied by the first weeding and replacement sowing, which must take place no more than a fortnight after sowing;
- The second weeding is combined with the final hoeing and ridging, which must be carried out four or five days after sowing;
- Weeding is carried out manually, but this is too difficult for farmers because of the high cost of herbicides.

[192] Interview with Deouyo, Koumra (Chad), 14 June 2017

[193] Groupe de travail Coopération Française sur les filières cotonnières en Afrique, p. 93.

[194] Interview with Beramgoto, Koumra, 08 June 2017

Photo 2: Weeding a cotton field after the young cotton plants have emerged and a cotton field in flowering.

Source: Nguehodjim photo, 16 June and Beyenan photo, 24 September 2017.

This photo shows the first weeding of the cotton plant followed by the removal of the seedlings to allow good growth and the flowering period. The cotton plant begins to bear fruit, which will lead to the opening of the seed cotton.

However, the NPKSB fertiliser (Nitrogen, Phosphorus, Potassium, Sulphur and Boron) must be applied to the cotton plants twenty days after the first weeding. This fertiliser "strengthens the plant and forty-five days later the nitrogen, which is urea, should be applied"[195] . Two (02) bags of NPKSB fertiliser are recommended for one hectare. The crop rotation recommended by the IRCT for crops grown on cleared land was to put cotton at the top of the rotation, followed by millet and then groundnuts, ending with as long a fallow as possible. Forty-five days is the date for the first treatment. The products are used according to the T.B.V (very low volume) technique. This involves spraying ten litres per hectare, mixing two sachets of concentrated products with nine litres of water[196] .

Micron Ulva+ equipment is used for spraying, and the operator must cover his head and fingers. The treatment must be carried out six or seven times, depending on the disease or pest observed. As far as possible, supervisors should determine the nature of the pests in the fields. For this reason, they are required to distribute the appropriate type of product to eliminate the attacks observed on the cotton plants[197] . The diseases and pests commonly encountered are :

- Aphids (Aphis gossypil) ;
- Melting seedlings ;
- Fritillary (lygus vosseleri),) caterpillars (helicoverpa armigera, Earias);
- Acariosis.

The treatment should be spaced a fortnight apart, while weeding as often as there is grass. Oxen should be used three times during the crop year. Experience has shown that cotton has a normal boll opening cycle of 120 days. So, if we take into account semi-early sowing from 1er June only, it should have an opening that can be harvested twice or even three times[198] . This is because early sowing allows the cotton plant to take advantage of the whole of the rainy season and the high temperatures in June; it reduces the damage caused by the major pest attacks in October and limits the risk of erosion of bare soil.

---

195 Interview with Beramgoto, Koumra, 08 June 2017

196 F. Nuttens, 2001, La production de coton graine en zone soudanienne, N'Djamena, Ministère de l'Agriculture, ONDR/DSN, p. 87.

R. Levrat, 1950, p. 135.

F. Nuttens, 2001, p. 123.

4) Duration of works and agricultural calendar

The duration of the work cannot be precisely estimated, as it depends on the difficulty of the work, which varies according to the nature of the terrain, the care taken with the plantation and the cultivation methods, with or without a horse and cart. After deducting the clearing time, it can be estimated at between one hundred and one hundred and twenty working days per hectare[199] , about half of which are in the rainy season, i.e. more than for millet.

Work is carried out according to a precise[41] calendar, with only a few variations. The cotton-growing season spans most of the year, from field preparation at the end of the dry season to harvesting, destruction of the cotton plants and marketing. The calendar drawn up by the IRCT's agronomists makes it possible to situate this work in time and thus to estimate its interference with the other activities of the growers.

However, the staking or demarcation of farmers' cotton fields within the group was carried out after the field survey. The table below illustrates the different stages in the agricultural calendar.

Table 2: Agricultural calendar in the Mandoul Oriental region

Month	Activities
1st April - 30th April	Choice of site
15 April - 15 May	Field staking, land clearing
15 May - 31 May	Distribution of seeds, spreading of manure, ploughing or hoeing of fields
1st June - 20th June	Sowing
20 June - 15 July	Replacement of missing items
1st July - 1st August	First weeding, loosening and spreading of oil cake or fertiliser, ridging
1 August - 20 August	Second weeding, ridging
1st September - 20th September	Third weeding, ridging
1 August - 30 September	Insecticide treatment
1st October - 31st October	Manufacture of drying racks, baskets, seed cotton and seed silos
15 October - 31 October	Cotton harvesting and sorting
1st November - 1st March	Cotton markets
1st January - 15th February	Uprooting and incineration of cotton plants

Source: CFDT, IRCT and ONDR

Preparing the field requires long, hard clearing work, which must begin in April. Only useful trees (acacia, shea, tamarind) and the stumps of medium or small-diameter trees, which will allow them to regrow at the end of the cultivation cycle, are left[200] . After burning the residues from clearing the land or from the previous harvest, the soil is hoed as soon as the first rains moisten it sufficiently to loosen it and allow the seeds to germinate properly.

The introduction of ploughing[201] made it possible to replace hoeing with ploughing. Ploughing was practised with an agricultural implement known as a "plough", which loosened and aerated the soil, ensuring good water retention on the surface and infiltration at depth. It requires more extensive clearing, but works better on heavy soils. Hoeing buries weeds and plant material, enriching the soil and reducing the need for initial weeding[202] . The main

---

Ibid.

200 J. Cabot, 1957, p. 95.

201 E.Vall et al. 2000, Études des pratiques et stratégies paysannes de traction animale dans les zones savanes cotonnières du Cameroun, Tchad et RCA, PRASAC/Irad, pp. 21-23.

202 *Ibid.*

benefit is that it saves the farmer time.

5) Harvesting the cotton and cleaning the stalks

The cotton harvest can start as early as mid-October, after the red millet harvest and during the white millet harvest: it is in full swing in November and December, and if necessary extends into January or even February, to allow all the bolls to burst. It's best to harvest[203] cotton when it is fully ripe, and then don't delay. Cotton is subject to wind, alternating dews and droughts, and an upsurge in aphids, so it gets dirty and deteriorates.

Harvesting cotton therefore requires two or three successive passes a few days or weeks apart[204] : the first, which is the most important in terms of quantity and quality, is reserved for white cotton. Subsequent passes require the cotton to be sorted to remove any yellow quality tainted by parasites, an operation that can be carried out during the harvest or at the hut. Harvesting is usually carried out by members of the family, although it is sometimes done collectively[205] .

Photo 3: Manual harvesting of seed cotton

Source: Photo by Yann Arthus Bertrand, 2005. Reproduction, Beyenan Ngarasndi

Work does not begin until the sun has dissipated the early morning humidity. The cotton is transported at the head of the cart, a heavy chore in which the whole family participates. It is first spread out for a few days on drying racks and then stored under cover, either in the attics or in the living quarters, or in earthen silos[206] .

As for the old cotton plants, they have to be cut flush with the ground at the end of the harvest to prevent regrowth, then piled up and burnt as a prophylactic measure. This operation, which still requires three days' work for one rope, has met with a great deal of resistance from the farmers. Most farmers did not "cut and burn them until the following year when they were

[203] E, Mbetid-Bessane et al. 2003, *Evolution des conditions de la production cotonnière en Afrique Centrale et ses conséquences sur les stratégies paysannes*, PRASAC, N'Djamena, p. 63.
[204] *Ibid.*
[205] R. Levrat, 1950, p. 215
[206] Interview with Beramgoto, Koumra, 08 June 2017

preparing their millet fields"[207] . It is therefore necessary to place particular emphasis on the development of cotton production and rainfall in the Mandoul Oriental region.

II - DEVELOPMENT OF SEED COTTON PRODUCTION IN THE EASTERN MANDOUL REGION

A - Development of seed cotton in tonnes on farmland

In the Mandoul region, cotton growing has been successful in some seasons, but has also declined due to natural and chemical factors. Between 1971 and 1975, the yield of grain cotton increased to 9,376 tonnes from an area of 2,513 ha. This drop in yield is explained by the delay in the agricultural calendar and the fact that few farmers were interested in growing cotton. The statistics for seed cotton produced between 1975 and 1980 are 5,793 tonnes from an area of 4,458 ha in the region[208] . This period coincided with the agricultural policy launched by President Ngarta Tom-Balbaye, which called for 75,000 tonnes of seed cotton[209] . However, the State encourages cotton production, which is vital for its financing. President Tombalbaye's political vision of the cultural revolution was a positive economic folly, as he asked Chadians to produce 75,000 tonnes of cotton in 1974[210] . For this reason, he declared that the priority objective 53 of the National Movement for Cultural and Social Revolution (MNRCS)[211] was to mobilise farmers to take an interest in seed cotton production. The 1980-1985 crop year produced a total of 7,233 tonnes of seed cotton from an average area of 6,124 ha[212] . Local people are interested in growing cotton because they need access to inputs to improve soil fertility and use them for food crops. It should be noted that COTONT- CHAD experienced crises[213] in the 1980s, as a result of which 65% of the local population abandoned this cash crop in favour of food crops.

From 1985 to 1990, the tonne of seed cotton increased to 9524 tonnes, covering an area of 6986 ha. These figures show the fall in the purchase price per kilogram; the Chadian government and COTONTCHAD have used new marketing strategies to raise the level of cotton production. In 1995, 11127 tonnes of seed cotton were grown on an area of 8934 ha[214] . This was the result of higher prices paid to producers for seed cotton. The 1995-2000 crop year saw a drop, with COTONTCHAD producing 11,285 tonnes from an area of 9,557 ha. From 2000 to 2005, a total of 1,733 tonnes were recorded, on a sown area of 1,125 ha[215] .

Cotton is grown on a large area of land in the region, but as a result of late rains, it is still in decline. Production of seed cotton in the 2005-2010 season was 13785 tonnes, corresponding to an area of 11987 ha. During these years, there was a good dose of rain, which helped to improve cotton yields. In the 2011-2012 season, we had 7,800 tonnes, or 1,800 ha of sown area. In 2013-2014, 14,142 tonnes of cotton were harvested from an area of 1,037 hectares.

---

[207] Interview with Beramgoto, Koumra, 08 June 2017

[208] D. Marambaye, 2002, "Évolution des conditions paysannes de production du coton au Sud du Tchad et ses conséquences sur les stratégies des paysans, Rapport de maitrise", PRASAC, N'Djamena, p. 63.

[209] R. Buijtenhuijs, 1965-1976, *Le FROLINAT et les révoltes populaires du Tchad*, Mouton, p. 175.

[210] Report by Chairman TOMBALBAYE, 27 August 1974, p. 69.

[211] Mouvement National pour la Révolution Culturelle et Sociale (MNRCS), created by TOMBALBAYE in 1973 after the dissolution of the PPT-RDA.

[212] D. Marambaye, 2002, p. 68.

[213] M. Fok, 2006, "Crises cotonnières en Afrique et problématique du soutien", *Biotechnologie, Agronomie, Société et environnement*, BASE [On line] volume 10, number 4, p. 312.

[214] . COTONTCHAD-SN, 2016, Rapport des activités agricoles, zone coton du Mandoul Oriental, pp. 912.

[215] *Ibid.*

From 2014 to 2015, COTONTCHAD (Koumra factory) saw remarkable growth, with 20590 tonnes of seed cotton recorded, representing 14345 ha of sown area. From 2016 to 2017, there were 27,000 tonnes over an area of 1,6887 ha[216] . The results of the agricultural campaigns take into account the constraints of nature, the areas sown according to the motivation of the producers. The graph below provides more information on the development of seed cotton production and the areas sown in Mandoul Oriental.

Graph1: Development of seed cotton in tonnes on agricultural land from 1971 to 2016

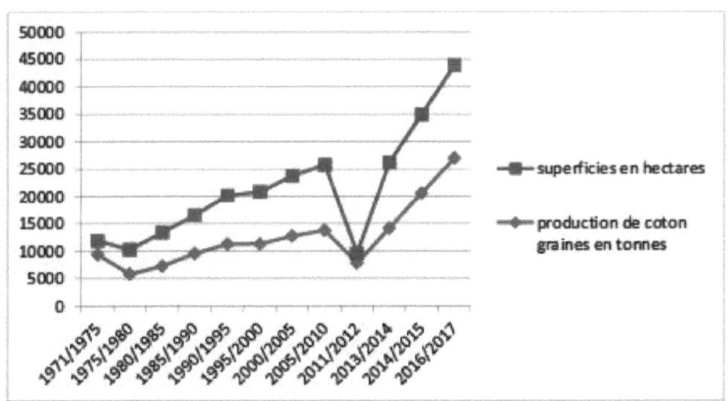

Source : COTONTCHAD-SN

B- Rainfall trends

Chadian agriculture is historically rain-fed, especially cotton growing, which is always waiting for the rains to arrive. This natural phenomenon is, in a way, the destiny of producers and even the rest of society. Rainfall patterns vary from crop year to crop year, sometimes decreasing, sometimes increasing, and this has very complicated effects on crop yields[217] . In the course of our research, we had to trace the data relating to the quantity of rainwater over the last 16 years in the Mandoul Oriental region[218] . The truth is that rain is a production condition that worries man, because he can no longer control it at the right moment. The following graph shows the evolution of rainfall in the Mandoul Oriental region.

[216] *Ibid.*

[217] A. Beauvilain, 1996, "La pluviométrie dans le bassin du lac Tchad. Travaux et Documents scientifiques du Tchad. Documents pour la recherche", Number V, CNAR, p. 89.

[218] COTONTCHAD-SN, 2015-2016, Rapport de l'antenne Radio sur la pluviométrie, p. 9.

Figure 3: Rainfall trends from 2000 to 2016

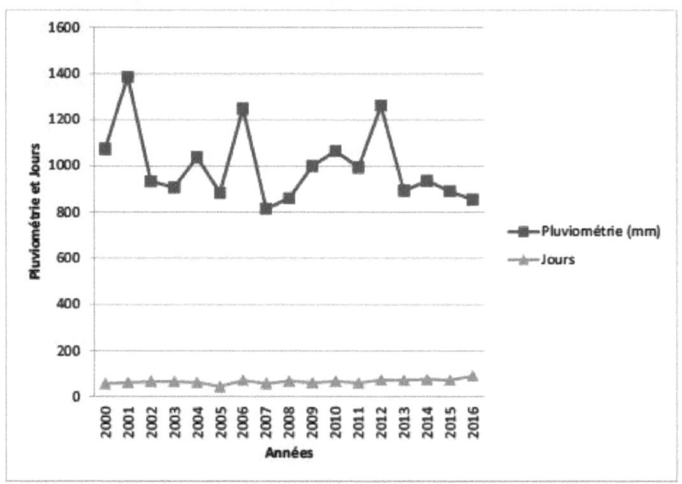

Source: Radio COTONTCHAD-SN (Koumra)

The right amount of rainfall varies between 800 and 1200 mm per year, which is good for crop yields.

III - THE CAUSES OF THE FALL IN PRODUCTION SEED COTTON

The main reason for the fall in production is the low yield in the field, reinforced by the reduction in the area sown.

A- Low yield

In the Mandoul Oriental region, agricultural yields have been erratic and on a downward trend, reaching a high level in the 2012/13 season compared with previous years. This decline can be explained by:

Soil exhaustion due to the over-exploitation of plots as a result of growing demographic pressure, the gradual elimination of fallow land, the unreasoned use of chemical fertilisers and the lack of restorative manure[219] . The fertilising elements exported each year for cotton cultivation are not sufficiently returned to the soil because the doses of fertiliser recommended in Chad are not respected, and are in fact below the norm due to the inadequacy of fertilisers and their high cost. However, an ITRAD study has shown that the recommended dose of 100 kg of NPKSB fertiliser plus 50 kg of urea per hectare is no longer profitable and should be revised to 150 kg/ha, plus the same amount of urea for spreading[220] . However, this new rate, which is lower than the rate of 250 kg of fertiliser per hectare (200 kg of NPKSB fertiliser plus 50 kg of urea) used elsewhere, is not yet widely used in rural areas.

Lack of organic fertiliser[221] , chemical fertilisers can only be effective on soils with a minimum organic matter content, but few growers bring this material to the field because it is unavailable, difficult to produce, difficult to work with and transport, and expensive.

---

[219] F. Nuttens, 2001, p. 47

[220] *Ibid.*

[221] *Ibid.*

The notorious lack of technical support for producers; the ONDR, which is responsible for cotton supervision, used to have a certain number of supervisory agents in the field. Experience in the field has shown that since the cotton crisis of the 1980s, the ONDR has shown little interest in supervising cotton growers and has therefore had few field agents. COTONTCHAD-SN, which should provide the local support desired by farmers and the Chadian government.

B- Decline in planted area

The fall in acreage immediately leads to a fall in production at the level of the co-tillers and at COTONTCHAD-SN[222] . It is often due to the discouragement of producers linked to the malfunctioning of the cotton company following the crises it has experienced, resulting in delays in the collection of cotton and the payment of income from seed cotton. This situation does not exclude the phenomenon of climate change in all its forms.

C- Climate change

The issue of climate change remains of concern to most African societies, including Chad. In 2001, Chad became aware of the phenomenon of climate change, with its vulnerability measured by a decrease or increase in rainfall, irregularity of rainfall and shorter rainy seasons interspersed with dry spells of varying length. The consequences of climate change for agriculture in Chad in general and in the Mandoul Oriental region in particular can be explained by extreme climatic events:

Floods, droughts, shorter rainy seasons, disruption of agricultural calendars and poor crop yields. However, it is remarkable that average annual rainfall has increased within the range of 800 to 1200 millimetres that characterises normal rainfall[223] .

Delayed rains have often disrupted the agricultural calendar, forcing growers to sow cotton outside the scheduled date, resulting in a drop in the area sown[224] . In the same vein, it should be noted that this is the case for dry spells or floods that occur at the height of the vegetative development period when the cotton plants have begun to bear fruit.

In short, cotton production requires a lot of work, with different methods for its development and a yield that varies according to production factors. Technical research carried out by agronomists has shown the importance of cotton growing in relation to the production system, in favour of power tillers for intensification. This system can be applied to food crops, respecting the sowing date and maintenance until maturity.

In Chad in general, and in the Mandoul Oriental in particular, the ONDR provides support to farmers for agricultural development, following the organisation of farmers and COTONTCHAD for the supply of inputs. In the wake of COTONTCHAD's financial crises, the ONDR has clearly withdrawn from the supervision and supply of agricultural equipment to farmers.

It is true that CO- TONTCHAD has made commitments to deploy cotton agents in the field, but the level of supervision still gives cause for concern in village associations.

[222] Interview with Deouyo, Koumra, 28 June 2017
[223] Interview with Ngar-rangaye, Koumra, 28 June 2017
[224] Interview with Ngar-rangaye, Koumra, 28 June 2017

CHAPTER III

COTTON MARKETING IN THE EASTERN MANDOUL REGION

Since its introduction to Chad, cotton has formed a market image between producers and cotton companies for its processing. As a result, southern Chad is the target area for cotton growing, contributing to the economic life of the country and its local population. Marketing cotton requires the players involved to be organised in order to make a profit. Long before the privatisation of the subsidiary cotton company in Chad, there were cotton market methods under the responsibility of village administrators. A few decades later, the cotton market system was modified under the name of "self-managed market"[225] where 95% of the responsibility is assumed by the farmers until the seed cotton is transported to the ginning plant for weighing at the Pont Bascule (PB)[226] . It should be pointed out that in Chad, the price of seed cotton on the internal market is set as a result of bipartite negotiations between COTONTCHAD-SN and the Local Coordination Committee, supervised by the State, with possible political arbitration concerning the granting of an additional subsidy to support the price of cotton.

The purchase price of seed cotton is then set by order of the Ministry of Trade. It is the same for the entire cotton-growing zone for one crop year. In view of the various interventions in the setting of the purchase price for seedcotton, farmers could not react, so they confined themselves to cotton production. From then on, the concept of seed cotton marketing consisted in analysing the itineraries of the self-managed markets at the level of the farmers and COTONTCHAD-SN.

I - MARKETING SEED COTTON

The seed cotton trade is the second stage in cotton production since its introduction in Africa in general and Chad in particular. The classic national market was managed by COTONTCHAD and was characterised by the presence of input distributors, payers, weighers, graders and seed cotton packers in the villages. As a result, it incorporates the organisation of farmers and COTONTCHAD.

A- The organisation of the seedcotton market at farmer level

Organising the cotton market is a highly technical activity that requires growers to adapt in order to avoid problems of all kinds once the seed cotton arrives at the ginning plant. In the Mandoul Oriental region, cotton growers have organised themselves in agreement with COTONTCHAD, through local coordinators at cantonal level, for the acquisition of inputs for cotton production through to harvest and purchase. The seed cotton market requires a number of channels for its operation, including the opening of a self-managed market, weighing and payment.

---

225 The farmers' request to the local coordinating committees to negotiate with COTONT- CHAD to revise the "charter of self-managed seedcotton markets", in particular article 16 - a request which, according to the farmers, has remained a dead letter to date, p. 5.

226 *Ibid.*

1) Opening up the self-managed market

Opening up the self-managed market is the first stage in commercial activity, which includes organising producers. It prepares their minds to be quicker in harvesting seed cotton exposed to animals or bush fires. To this end, the president of the village association (AV) invites the growers to clean the buying centre (CA) and then leave with their stock of seedcotton[227] . Various means of transporting the cotton were used by the farmers, some of whom were forced to move the cotton to the buying centre.

However, to keep costs down, they force their families to fill traditional baskets, containers and bags with fertiliser in order to transport the cotton on their heads to the market, no matter how far away it is. Otherwise, they have to hire carts and oxen for transport. Farmers who had the means of transport could use them at any time to take their stock of seed cotton to the buying centre. The work involved was a real chore for the farmers, without measuring the negative consequences[228] . When the seed cotton was purchased, the farmers contributed their millet reserves to pay the labourers, even though COTONTCHAD had to pay them on the basis of the tonnes of cotton shipped in the poly ben box.

Added to this is the average quality of production, which can be improved by creating homogeneous batches of seed cotton, grouping together harvests from plots that have received the same technical itinerary and separating the first harvest from the next. An area measuring at least 50 x 75 metres is needed, free of trees, cleaned and swept clean, and a shelter must be built, including a shed for the purchasing team: 4 to 5 metres long, 3 to 4 metres wide and at least 2.5 metres high[229] . After this exercise, the inspector intervenes to monitor batches of seedcotton stored before the weighing is authorised.

Photo 4: Self-managed seed cotton market

Source: Didrot Nguepjouo, 2011, p. 22. Reproduction: Beyenan Ngarasndi

This photo shows the purchasing centre (CA) for which the self-managed market took place in the presence of the control agents before the weighing.

2) The role of the control officer at the seed cotton purchasing centre

The role of the inspector at the seed cotton market is to ask producers to be regular: to sort the

[227] Interview with Madrangaye, Koumra, 06 June 2017

[228] J. Lhuilier, 1900-1950, "Tchad, le coton", Tropique, no. 328, p. 120.

[229] M. Cretenet, 2006, *Le guide technique de la production du coton graine de qualité*, Paris, L'Harmattan, pp. 101-109.

cotton and keep it in the appropriate place before selling it.

- Storage and sorting of seed cotton: at farm level, the harvested seed cotton must be dried on racks on the day of harvest and then sorted according to harvest number and visual quality into three categories:

First choice: white cotton, sorted, dry, free of spotted fibres, immature boll lodges, stem residues, bacteria and other impurities from whole unopened or opened bolls.

- The second choice: unsorted white cotton or clean spotted cotton, free of immature boll lodges, stem residues, bacteria and other impurities, whole unopened or opened bolls.
- The third choice: cotton made up of the sorting residue, it is strongly coloured, generally immature and dirty[230] .

However, the official grading of seedcotton according to these criteria is carried out at the village association buying market. This is done visually using a standard seed cotton box with three compartments. The end boxes are used as references and contain samples of seed cotton representative of the batch of cotton to be graded. Grading is carried out by comparison with the references whose cotton batch is graded according to whether the appearance of the sample is closer to the first or second reference, but it is ranked third. Once the sample has been checked, the farmers are expected to bring their bales to the scales. The weighing team must record the references of the bales weighed, their weight and origin, the variety and the generation of multiplication in an appropriate notebook. The seller must sign his name in front of this notebook and take with him the ticket showing all the references.

The first-grade seed cotton is packaged, placed in the lorry and transported first to the ginning plant. In the crate to be filled, the tarpaulins are positioned on the ground to recover the cotton that has fallen with as little impurity as possible[231] . The seed cotton is placed in the box to save as much space as possible. If the different qualities are to be loaded onto the same lorry, care must be taken to separate them. Each box must have a sign, which is the black part, painted with chalk to inform the factory of the origin of the seedcotton (village, buying market), the variety and propagation generation, the approximate weight of the seedcotton loaded and its commercial category[232] .

However, the self-managed markets (SMM) are a primary seed cotton marketing activity managed by the village associations and governed by the charter that sets out the organisational procedures. After the buying team has made a summary, two conveyors are responsible for transporting the filled seedcotton in the lorry to the ginning plant, via the weighbridge, where it is weighed to determine the weight of the village and that of the factory. The difference remains either in surplus or in deficit in the account of the VA concerned. To do this, COTONTCHAD is obliged to buy all the cotton produced by a regularly constituted VA. In the event of default, i.e. failure to comply with the charter or unpaid inputs, the VA's market may be suspended.

3) Payment to farmers and repayment of CO- debts TONTCHAD

Payment of the VAs was supposed to take place just after the end of the self-managed market. "At the ginning plant, COTONTCHAD takes the money from the equipment and inputs on credit before paying the VAs concerned"[233] . The VAs will in turn recapitalise to satisfy their

[230] COTONTCHAD , Guide pratique des activités commerciales de coton graine, p. 11.
[231] *Ibid.*
[232] Interview with Deouyo, Koumra, 14 June 2017
[233] Interview with Deouyo, Koumra, 14 June 2017

members, which is no easy task given the joint and several guarantee system in the groups. This operation only concerns growers who have not been able to repay their debts for inputs and agricultural equipment delivered by CO- TONTCHAD. VA managers use a number of strategies to cover debts before payment is made: the most common is to charge the debt to the group, which is equivalent to recovering it directly from the farmer concerned and his family, which can go as far as seizing agricultural assets or available livestock[234] .

In fact, the fact that producers are unable to repay input credit is due to a number of different factors:

> Production devastated by regional climatic events (flooding, drought);

> Production shortfalls linked to shortcomings directly attributable to COTONCTHAD in the management of productivity campaigns (delays in the introduction of inputs, for example);

> Financial losses due to VA practices: transfer of the delivered input to other production, delivery of cotton under cover of another VA, group stoppage of cotton production after receipt of inputs and transfer and sale of stock.

Once the debts had been recovered, the AV presidents could continue with the payments to the growers, taking into account the order of the recorded weighing. It should be noted that for some AVs, the surpluses and rebates granted by CO- TONTCHAD are intended for social work. However, many have "used this money between the purchasing team and the village chiefs for their own needs"[235] . This is what often leads to conflicts in the VAs, as the activities of the self-managed market are the result of the methods used to set seedcotton prices.

B - Mechanisms for setting seed cotton purchase prices.

The mechanisms for setting seed cotton prices for growers are the responsibility of the parastatal cotton companies through a system of regulated prices. These companies have a monopoly on the supply of inputs and agricultural equipment to producers[236] . As a result, producers have little influence over the types and quantities of seed they use. On the face of it, the parastatal companies control purchases of seed cotton and most of the ginning and marketing of cotton fibre, both internally and externally, as well as seed cotton co-products. However, it should be pointed out that there is a single purchase price for seed cotton set in all the cotton-growing zones of Chad[237] . This price may be changed or maintained during the following crop year.

The current price-fixing system offers a degree of stability to producers, but at a relatively high cost in terms of the agricultural equipment granted by the bodies. In addition, as part of the operation of the Caisse de Stabilité des Prix du Coton (CSPC), a cotton price-fixing mechanism was set up in 1968, which guarantees and stabilises the price of cotton to producers from the time COTONTCHAD began operating the sector[238] . To this end, the Government sets the price of cotton to be paid to growers, which is later determined by applying a mechanism whose calculation formula is equal to 17% of the price of fibre less taxes and other export costs. If the price paid to growers is higher than the calculated price,

[234] Interview with Deouyo, Koumra, 14 June 2017

[235] Interview with Djimako, Ngomanan, 18 May 2017

[236] Macra Tadin, 1983, "L'Intervention de l'Etat dans le secteur cotonnier au Tchad", Doctoral thesis in Public Law, University of Toulouse, p. 207.

[237] Fauba Padache, 2010, *Étude du mécanisme de fixation du prix de coton graine au Tchad*, Bamako, Mali, p. 45.

[238] *Ibid.*

the CSPC will reimburse COTONTCHAD for 80% of the losses incurred. If, on the other hand, the price is lower, COTONTCHAD pays CSPC 80% of its profits, less taxes and depreciation[239] . This pricing policy guarantees income for producers, encourages them to produce more cotton and thus provides foreign currency for the State.

1) Chadian government policy on the mechanism for setting seed cotton prices

The Chadian State plays the role of observer with regard to the application of the mechanism for setting seed cotton prices, which is the responsibility of the Ministry of Trade and the Ministry of Agriculture at the time of negotiation of the first payment per kilogram of cotton in partnership between the producers and COTONTCHAD (Joint Committee). It finances the positive difference determined by the mechanism and the payment previously received by the producers. Thus, the Chadian government's policy for setting the price of cotton to producers consists of committing to financing through balancing subsidies granted to COTONTCHAD[240] . Since the 2001/2002 season, the State's total subsidies to COTONTCHAD amount to 73047 million francs, of which approximately 2/5 (28500 million) are negative rebates. The balancing subsidies also take into account the State's share of input approach costs, which since 2005/2006 have totalled around 5069 million francs[241] .

State subsidies to COTONTCHAD take several forms depending on the year. The agricultural policy adopted since 2001/2002 has been renewed in the draft performance contract between the State, COTONCTHAD, the Institut Tchadienne de Recherche Agronomique pour le Développement (ITRAD) and the Office National de Développement Rural (ONDR)[242] . This draft contract, which is supposed to cover the period from the 2007/2008 to 2009/2010 marketing years, contains other government commitments relating to the cotton sector.

Although the contract is still at the draft stage, the State has implemented some of its terms: the inclusion in the 2008, 2009 and 2010 budgets of subsidies totalling 4,012 million CFA francs for the ONDR and 1,827 million CFA francs for the ITRAD. The ONDR received a state subsidy of 600 million CFA francs to cover the cost of recruiting field staff, purchasing rolling stock and running costs. In relation to the cotton sector, there is a Memorandum of Understanding signed on 05 February 2009 between the ONDR and COTONTCHAD, the aim of which is to determine the terms of the partnership between the two institutions for the monitoring of 2008/2009 seed cotton marketing activities[243] . There is also a partnership framework agreement[244] signed in July 2009 between ITRAD and COTONTCHAD, the aim of which is to strengthen the partnership between research and the cotton company by formalising the framework for the provision of services and expertise to COTONTCHAD.

2) The price-setting mechanism for seed cotton proposed by SOFRECO

21

The mechanisms for setting seedcotton prices proposed by SOFRECO[245] (Société Française

[239] Marca Tadin, 1983, p. 211.

[240] C. Araujo-Bonjan and J.F. Brun, 2000, "Are cotton price stabilisation policies in franc zone Africa doomed?" Revised version of a paper presented at the conference *"Dynamique des prix et des marches de matières premières : analyse et prévision"*, http://www.agriculture.gouv.fr/spip/IMG/pdf/coton_-marche.pdf, accessed, 14 July 2017

[241] Macra Tadin, 1983, "L'intervention de l'État dans le secteur cotonnier au Tchad", Doctoral thesis in Public Law, University of Toulouse, p. 98.

[242] *Ibid.*

[243] C. Araujo-Bonjan and J.F. Brun, 2000, pp. 17-19.

[244] *Ibid.*

[245] Société Française de Réalisation, d'Étude et de Conseil (SOFRECO), 1996, "Étude d'un mécanisme pour la Détermination du prix d'Achat du coton graine aux producteurs", p. 56.

de Réalisation d'Étude et de Conseil) are based on the following three principles, given the fundamental difficulties in determining a main criterion for setting the grower's remuneration in relation to that of other operators.

- The remuneration of each partner must be calculated independently in accordance with the business logic as the basic principle retained in the Shareholders' Pact[246] . The solution which consists of aggregating the results of the sector and distributing the profits between the main players (producers, cotton company and the State) has disadvantages[247] . Producers' remuneration must be calculated on other bases, independently of the sector's results;
- The price of seed cotton to the grower must be calculated on the basis of a fixed percentage of the price of the sector in CIF position. This percentage is defined as the ratio between the producers' cost price and the total cost price of the fibre;
- The cost prices taken into account in the calculations are objective cost prices that can be achieved in the short term with realistic efforts on the part of those concerned, and not actual cost prices.

The cost price proposed as the producers' objective is that of a producer using the F2 productivity formula, including 2 bags of NPKSB and 1 bag of urea per hectare, and ploughed cultivation. However, the selling price of the fibre in the CIF position is outside the average price of COTONTCHAD sales: the arithmetic average of the Cotton Outlook (Cotlook) A index over 7 months, from January to June, corresponding to the period during which the fibre is sold by COTONTCHAD. As a result, the implementation of the mechanism is constrained by the existence of a time lag between the moment when the price is indicated to producers in April before sowing and the moment when the calculation parameters are known in June of the following year. This shows the cyclical variation in world prices for cotton fibre and therefore in the purchase price to growers, which, at certain phases of the cycle, risks turning them away from cotton growing. To this end, SOFRECO has proposed four solutions for implementing the mechanism, two of which better meet the expectations of the Comité de Réflexion et de Suivi de la Filière Coton (CRSFC)[248] :

- Payment in two instalments. The price is calculated in two stages: on the basis of the forecast sales price for the sector and the tonnage of seedcotton, the forecast price which will be paid to growers at the time of marketing is calculated and announced to growers before sowing; the final price is calculated on the basis of the actual figures for the season. This second calculation is used to determine production at the end of the season. The solution allows added value to be shared fairly between the three main players in the sector, but makes no provision for the event of a fall in prices;
- Payment in two instalments but with capping. In line with the previous solution, this differs in that the second instalment is paid in full only if it falls within a pre-determined price range, in agreement with the producers.

Otherwise, the surplus is deposited in a reserve account. If the price is higher than the range, it will be borne by the reserve within the limit of the sums available or lower than the range. This solution offers many more advantages than the previous one, because of its ability to absorb sharp variations in producer purchase prices. It also enables COTONTCHAD to better

[246] *Ibid.*
[247] *Ibid.* p. 58.
[248] The CRSFC expects a mechanism that would include a floor price when seed cotton is marketed and a subsequent premium that would enable the three partners in the sector to share the added value fairly, while taking account of changes in world cotton prices.

control its production costs, since the "purchase of seed cotton" item will represent a percentage that the principle of the reserve depends on the producers' support[249].
Furthermore, the increase in the mechanism for buying seedcotton from growers must not be manipulated by any of the parties: COTONTCHAD or the growers, and the risks of a fall in prices must be better shared between upstream and downstream players. In theory, however, growers should be paid on the basis of the objective parameters of COTONTCHAD's cost price and its commercial policy[250].

C - Purchase price trends for seed cotton from 1971 to 2016

Buying prices for seed cotton in Chad have fluctuated from crop year to crop year in relation to world market prices and the cost of transporting cotton fibre from the cotton-growing areas to the port of Douala (Cameroon). However, cotton purchase prices are set by a joint committee, with state intervention through subsidies between the parties. Between 1971/1972, the purchase price of seed cotton was set at 28 FCFA/kg, then 43 FCFA/kg between 1974/1975. It then rose gradually, reaching 100 FCFA/kg in 1988/1989. This price was reduced to 90 FCFA/kg between 1989/1990, where it remained until 1993/1994. This was justified by the year in which the CSPC was dissolved, and rose to more than 100 FCFA/kg in the 1994/1996 season, to 120 FCFA/kg and 170 FCFA/kg in 1996/1997[251]. As part of the structural adjustment programme, a new price-setting mechanism was introduced in 1997 at the behest of the World Bank (WB)[252]. This mechanism is determined on the basis of 19.3% of the average world fibre price measured by the Cotlook A index[253]. It has operated for 14 marketing years, from 1997/1998 to 2009/2010. However, this mechanism requires COTONTCHAD to pay a positive price differential to producers if the price paid in advance is lower than the calculated price. Otherwise, COTONTCHAD runs a risk of loss. This agreement was signed between the joint committee and the State under Order No. 002/MICA/DG/01 of 09 April 2001 specifying the variation in seed cotton prices between 150 and 194 FCFA/kg over the 14 seasons. It should be noted that in 2011, the mechanism was declared obsolete by the main players in the sector. Since then, the price has been set without reference, often by the State. The State set the price of seed cotton at 215 FCFA/kg between 2011/2012 and 2012/2013, and increased it to 240 FCFA/kg in the 2013/2014 and 2015/2016 seasons[254]. The table below shows changes in the purchase price of seed cotton from 1971 to 2016.

Table 3: Change in seed cotton purchase prices in Chad from 1971 to 2016

Years	Price (FCFA/Kg)	Years	Price (FCFA/Kg)
1971	28	1994	120
1972	29	1995	135
1973	31	1996	170
1974	41	1997	194
1975	75	1998	178

[249] C. Araujo-Bonjan and J. F. Brun, 2000, p. 24.
[250] *Ibid.*
[251] COTONTCHAD, 2000, Rapport des campagnes agricoles et commerciales, usine d'égrenage de Kou- mra, p. 8.
[252] World Bank, 1998, Politiques cotonnières en Afrique francophone, problématiques (version préliminaire), Washington: Word bank, p. 58.
[253] E. Gerald, 2006, "Le marché mondial du coton : Évolution et perspectives", *in Cahiers Agricultures*, issue 1, p. 19.
[254] Interview with Pafing Chiakré, Koumra, 12 June 2017

1976	45	1999	150
1977	50	2000	150
1978	50	2001	165
1979	50	2002	190
1980	50	2003	190
1981	60	2004	190
1982	70	2005	190
1983	80	2006	160
1984	100	2007	160
1985	100	2008	160
1986	100	2009	180
1987	100	2010	180
1988	100	2011	180
1989	90	2012	215
1990	100	2013	240
1991	90	2014	240
1993	90	2015	240
1994	90	2016	240

Source: COTONTCHAD-SN and Chad Ministry of Agriculture.

Trends in seed cotton prices in Chad from 1971 to 2016 have been up and down, reflecting the various factors affecting cotton growers during the growing season. These include natural disasters, poor seed varieties, poor quality agricultural inputs and the discouragement of farmers in the commercial system. The reasons for this often have to do with the price of agricultural inputs and equipment.

1) Changes in the price of agricultural inputs and equipment

Input prices are set by the technical committee and communicated to growers at the start of the crop year. They are the same for the whole of Chad's cotton-growing zone. The prices of agricultural inputs and equipment are shown in the table below for the period 1995 to 2016.

Table 4: Changes in prices of agricultural inputs and equipment from 1995 to 2016

Years	NPKSB 50 Kg bag (F CFA)	UREE Bag 50 Kg (F CFA)	INSECTICIDE Dose 12.5 ml (F CFA)	BATTERY (F CFA)	PULVERIZER (F CFA)
1995/1996	13585	9000	1680	120	39525
1996/1997	17065	10775	1680	175	40215
1997/1998	18295	14975	1445	160	40215
1998/1999	16200	12340	1440	135	40215
1999/2000	14976	12340	1018	135	27851
2000/2001	13995	13423	1129	154	27851
2001/2002	14719	13702	983	135	27851
2002/2003	14370	13702	950	139	29694
2003/2004	14986	13120	983	149	29694
2004/2005	15028	13842	770	140	27500
2005/2006	14443	13666	844	140	27500
2006/2007	17023	16256	855	135	23541

2007/2008	15026	14779	855	135	23541
2008/2009	15026	14779	844	140	27500
2009/2010	15000	14000	850	100	23003
2010/2011	15000	14000	850	100	23003
2011/2012	15000	14000	850	100	23003
2012/2013	15000	14000	850	100	23003
2013/2014	16000	15000	900	100	25000
2014/2015	16000	15000	900	100	25000
2015/2016	16000	15000	900	100	25000
2016/2017	16000	15000	900	100	25000

Source: COTONTCHAD-SN, Production Department

2) How growers acquire inputs and collect debts

The way inputs are acquired can be summed up as follows: the cotton company supplies inputs (seeds, fertilisers and insecticides) to cotton growers on credit. The *sine qua non* for any cotton grower to benefit from these inputs on credit is to belong to a joint guarantee group formed by the farmers, known as the Village Association, represented by the sub-groups[255] . The latter are grassroots organisations of cotton growers under the control of an Agent Cotonnier de Terrain (ACT). COTONTCHAD is responsible for delivering the inputs to the growers, but if it is impossible to do so due to poor road conditions, or if the requirement expressed by the growers is for a small quantity, the VA concerned must make every effort to have its orders moved. Once the agricultural inputs were in place, the village associations looked for strategies to recover the debts owed to COTONT- CHAD.

3) The method of collecting debts or repaying loans

Repayment of input credits granted to growers is made when the seedcotton is marketed[256] . The cotton company directly withdraws the amount of the credit when the total value of the seedcotton transported by the VA to the factory is paid. This strategy enables the cotton company to prevent certain producers with low production, who cannot cover the amount of credit granted, from refusing to repay their credit. As a result, the solidarity guarantee rules prevail in the VAs. Producers in a deficit situation vis-à-vis their organisation in terms of loan repayments are obliged to repay to the group the value of their loan covered by the solidarity guarantee. Refusal to repay the loan meant that these producers were simply excluded from the group[257] .

II - ORGANISATION OF COTONTCHAD IN RELATION TO THE PURCHASE AND PROCESSING OF SEED COTTON

COTONTCHAD's organisation for the purchase of seed cotton is based on the collection of cotton from the VAs, the transport of seed cotton to the ginning plant, the transport of lint to the outside world after processing and the management of co-products.

A - Purchase of seed cotton by COTONTCHAD

COTONTCHAD's organisation waits for the right moment to buy the seed cotton from the VAs, either directly from the poly ben or by stocking it at the buying centre before the lorry arrives to take it on board. In order to meet producers' needs in terms of the quality of

255 COTONTCHAD-SN, 2008-2009, Rapport d'activités de production et commerciale, p. 6.

256 *Ibid*, p. 9

257 Interview with Mbang Adoumbe, Nderguigui, 21 June 2017

seedcotton produced, it grants credit for inputs and agricultural equipment to the VAs. It has to be said that transport plays an important role in the collection of seedcotton, which deserves to be analysed.

1) Transport of seed cotton by COTONTCHAD to the ginning plant

The commercial activity of seed cotton has to follow an accumulation process where COTONTCHAD has to "place the inputs and then transport the cotton from the origin to the ginning plant"[258] . On arrival at the factory, the lorry must pass through the weighbridge to weigh and check the seed cotton for classification before processing. The seedcotton is therefore weighed at the ginning plant at two levels: single weighing on entering the factory and double weighing after unloading.

For simple weighing, parameters are known by the vehicles and containers and stored in advance in the computer. This weighing automatically gives the net weight of the seed cotton. Reference tickets are available showing the net weight, gross weight and tare. The COTONTCHAD factory in Koumra does not have enough lorries for transhipment.

The weighing of private vehicles is carried out in the same way as for COTONTCHAD. Private vehicles are weighed in the same way as COTONTCHAD vehicles. Weighing at the weighbridge before entering the factory gives the gross weight of the seedcotton, and weighing on leaving the factory indicates the tare weight of the vehicle. The net weight of the weighed seed cotton is obtained by the difference: N = B - T[259] .

The printer displays the weighing results (tickets). After this, two display screens simultaneously communicate the weighing results to the operators in the computer room and to the VA representatives in the waiting room. From this point onwards, transparency in the weighing process at the ginning plant was confirmed, as many conveyors were suspicious of the weighbridge operators because they were not familiar with the computer programming the weighing of seedcotton[260] .

In addition, when the weighing ticket is issued, it takes into account the contents of the transport slip which the driver must hold in his hand and which is relevant to the fuel or lubricant coupon. The latter also takes into account the mileage travelled (the vehicle's departure and arrival counters) and the fuel consumption based on the quantity on arrival compared with the quantity on departure, in order to determine the total consumption per kilometre. This exercise includes the weighing code, the material entry, the vehicle code and the identity of the driver[261] . Following the transport and weighing stages, the seed cotton must be checked for classification and placed in the feeder for transfer to the ginning mill.

2) Production of baled cotton fibre

The ginning of seed cotton enables the cotton seed to be separated from the fibre and at the same time the cotton waste to be separated. Ginning is the main activity of COTONTCHAD.

The cotton is fed into an inverted triangular tray containing a rod fitted with helical blades that push the product towards the gin. The gins are fitted with saws, brushes and pins placed in a chain. This chain work is carried out by means of belts linked to pulleys, making it possible to move the brushes and saws, which are fixed to the cylinders inside the gin machines, at the exit of the chestnut mechanism, from which the product separates into two parts: the seeds on one side and the fibres on the other. The fibres fall into a box which is then driven for compaction by a cylindrical rod with

[258] Interview with Deouyo, Koumra, 16 June 2017
[259] Interview with Deouyo, Koumra, 16 June 2017
[260] Interview with Deouyo, Koumra, 16 June 2017
[261] Wawe Haroun, 2015, "Pratique de contrôle budgétaire, COTONTCHAD-SN", Professional Master's thesis, University of Ngaoundéré, p. 67.

helical blades. If the cotton has to be ginned in excess and the ginning machine is delayed, the surplus is transferred to a reserve chamber and the product is sucked up again to be ginned[262].

The gin is fitted with windows containing scrapers to remove impurities and rods to remove dirt. The fibre sucked in is then moistened by the steam and the condenser condenses it. In this case, the steam is produced by the fire following the burning of the oil and is pushed by a high-pressure fan towards the box from which the dry fibre is displaced in order to make it damp. However, the cotton circuit feeding the machines must be in harmony with the bale-wrapping system. In fact, it is a "job that requires speed on the part of the staff to avoid accidents, especially when the bales of cotton fibre are being prepared inside the machine and taken out"[263]. Wrapping is the final stage in cotton processing at the ginning plant.

Photo 5: Cotton ginning at Koumra.

Source : cliché Djerabe, 14-06-2017

The cotton gin separates the seed cotton from the fibre to produce a finished product for export.

The conditioning system consists of a general condenser, a "cotton fibre chute, a cotton fibre feeder, a tamper, a baler, systems for binding and covering the bales, and bale transport systems"[264]. The baler consists of a frame, one or more hydraulic cylinders and a hydraulic circuit. The strapping sub-systems can be fully manual, due to its semi-automatic position, or fully automated. The packaging ties are generally steel rods or flat steel or plastic straps. Six to ten ties are usually placed along the length of the bale, although a 'continuous spiral tie' is sometimes used[265]. Once the bale has left the baler, the pressure exerted on the ties depends on the uniformity of the fibre distribution, the weight of the bale, its dimensions, the density at which the bale has been pressed, the moisture content, the length of the ties and other factors. Ties must be adapted to the baler to avoid contamination and handling problems. To prevent deterioration of the fibre in the bale, the moisture content of the cotton in the bale must not exceed 7.5% at any point. Fibre deteriorates considerably more with increasing

---

Interview with Djimtolabaye, Koumra, 14 June 2017

Interview with Djimtolabaye, Koumra, 14 June 2017

264 Interview with Djimtolabaye, Koumra, 14 June 2017

265 Interview with Ngaryedji, Koumra, 14 June 2017

moisture content, particularly above 9%[266] . The bales must be completely covered, including the openings made for sampling, and the bale covering must be clean, in good condition and strong enough to adequately protect the cotton. In the case of outdoor storage, the packaging must contain ultraviolet inhibitors depending on the expected duration of storage. The table below illustrates the quantity of cotton fibre produced by the Koumra ginning plant.

Table 5: Cotton fibre production in tonnes, 1970-2016.

Production	Tons
1970/1980	27000
1980/1985	42127
1985/1990	77856
1990/1998	94000
1998/2005	12900
2005/2010	33700
2010/2016	98000

Source: COTNTCHAD-SN, Production Department

This table shows the evolution of cotton fibre production according to the crop year. The trend in this production is due to different circumstances depending on the motivation of seed cotton producers, which has a direct influence on the tonnes of cotton fibre during processing. This can also be explained by the phenomena mentioned above.

Photo 6: Preparation of the cotton lint bale in the machine and its removal.

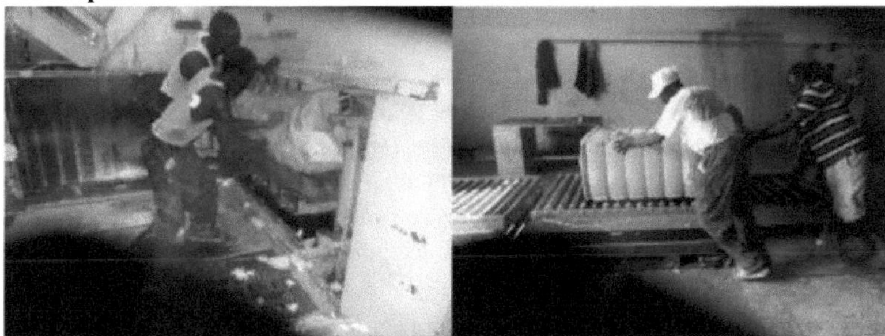

Source : Cliché Beyenan Ngarasndi, 14-06-2017

The bales produced are weighed, numbered and sampled. The cotton fibre production department has estimated the time taken for the bale to exit the machine at around 4, 5 and 6 minutes[267] . It should be remembered that there are several different qualities of cotton fibre, whose labelled marks give the following indices: 34.50; 32.50; 30.60; 36.50, as well as their weights, which vary between 224 kg, 234 kg, 243 kg, 250 kg and 265kg[268] . This being said, the weight of the cotton fibre depends on whether it was sown early or late, and also on the seed variety. Consequently, the numbering of bales of cotton fibre and samples determine qualities and prices on world markets.

3) Transporting cotton fibre to the outside world

The transport of cotton fibre follows almost the same procedure as that of seed cotton, where

[266] Interview with Djimtolabaye, Koumra, 14 June 2017

[267] Interview with Adamou, Koumra, 14 June 2017

[268] Interview with Marmai, Koumra, 14 June 2017

the quality of the cotton fibre must be checked by means of samples. After the bales have been loaded, the vehicle must be weighed at the weighbridge before being transported by road via Ngaoundéré to the port of Douala (Cameroon)[269] . Bales of cotton fibre are transported by COTONTCHAD-SN lorries and private vehicles. In this respect, COTONTCHAD-SN's transport prices and those of the private sector differ considerably in terms of the distance covered. In this way, data such as the cost of transport provides some indication of the difficulties of transport routes in Chad. It is difficult to propose a synthetic indicator representative of the level of transport prices for cotton fibre in southern Chad, and more particularly in the Mandoul Oriental region. There are various reasons for the high cost of transport: the long distance, the high cost of fuel at any given time, and the road charges, which consist mainly of taxes levied by all customs and police agents[270] .

Photo 7: Storage and loading of cotton fibre bales, Koumra factory.

Source: photo by Beyenan Ngarasndi, 14-06-2017

Bales of loaded cotton fibre must be traded externally. Poor quality cotton fibre is sold within the country for the manufacture of mattresses, and the seeds separated from the fibre are crushed and sent to the Huilerie et Savonnerie (HS) in Moundou. The ginning plant at Koumra uses the cotton hull as firewood after hulling[271] . This produces the steam that turns the turbine, providing the electrical energy that powers the company's various departments. From then on, the by-products are evaluated according to the daily ginning capacity and during a campaign to obtain tonnage. Generally, the hulling workshop evaluates 70 to 80 tonnes[272] of cottonseed per day. The production of cottonseed oil and cake should therefore be analysed.

4) Oil and cake production

The production of oil and cake is the result of the quantity of cottonseed produced annually. This production has not evolved in the same way since the creation of the oil and soap factory in partnership with COTONTCHAD. The table below shows the tonnes of oil and cake produced from 2000 to 2016.

Table 6: Oil and cake production in tonnes

Campaigns	Neutral cottonseed oil		Refined oil		Oil cake in tonnes
	200-litre drums	20-litre cans	tonne	Cardboard	

[269] Interview with Telembaye, Koumra, 13 June 2017
[270] Interview with Telembaye, Koumra, 13 June 2017
[271] Interview with Vaitchiou Vaibra, Koumra, 20 June 2017
[272] Interview with Vaitchiou Vaibra, Koumra, 20 June 2017

						15 litres	
	(tonnes)	Number Futs	(Tonnes)	Cans			
2000/2001	3695	20005			1176	121984	16936
2001/2002	6648	36442			2377	199037	24119
2002/2003	5524	30174			1731	126436	25794
2003/2004	3869	21144			830	60607	18448
2004/2005	1736	9515			250	9693	12682
2005/2006	1474	8052	1290	70450	37	1391	14603
2006/2007	2376	13025	1121	61461	24	1225	15112
2007/2008	1602	8781	858	47025	475	34221	10191
2008/2009	1003	5500	3554	194780	257	15713	12777
2009/2010	9785	4767	5462	19397	353	18987	11963
2010/1011	8857	3879	4981	27123	482	20627	14893
2011/2012	6998	4983	5927	3440	636	21284	13987
2012/2013	3987	5623	6128	2319	534	20123	13624
2013/2014	4225	3920	5810	1980	576	1996	12975
2014/2015	3194	2975	4914	2123	595	1897	1368
2015/2016	5220	2730	4168	2787	476	20133	15265
2016/2017	5724	3762	3947	3811	439	19217	15142

Source: Oils and Soaps Division (DHS).

a) Managing cottonseed oil and cake

The oil and cottonseed cake market is managed by COTONTCHAD-SN's Direction de l'Huilerie Savonnerie (DHS), which is the only producer of cottonseed oil and cottonseed cake on the Chadian market. The oils produced are: neutral oil packaged in drums, and refined oil packaged in one-litre plastic bottles. Cottonseed cake is packaged in 70kg bags. The supply of cottonseed oil depends on household needs, but does not really meet demand.

> Cottonseed oil supply: the quantities of cottonseed oil produced depend directly on variations in seed cotton production, which is not stable and has been falling steadily for several years. Production of neutral cottonseed oil by the DHS varied from 7,268 tonnes in 2000/01 to 4,660 tonnes in 2008/09, a drop of around 40%. The average annual production for the last nine campaigns is 5,644 tonnes. Refined oil production fell from 1,176 tonnes in 2000/01 to 257 tonnes in 2008/09, a drop of 78%[273] . This average annual production of cottonseed oil corresponds to around 5 million litres, compared with a theoretical capacity of 18 million litres per year. It is clearly insufficient to satisfy domestic demand. The supply of oil is limited by the performance of the production tool and by dependence on sources of raw material supply (COTONTCHAD-SN seed cotton).

> Cottonseed cake supply: DHS is the only industrial entity producing cottonseed cake in Chad. The second largest cottonseed cake producer in the sub-region is the SODECOTON oil mills in Garoua and Maroua in Cameroon. But Garoua's cottonseed cake is mixed with hulls and is therefore less rich in protein than Chad's.

Annual meal production varied between 12,682 tonnes in 2004/05 and 25,794 tonnes in 2002/03. Average annual production was less than 17,000 tonnes between 2000/01 and

[273] Direction Huilerie Savonnerie, 2009, Production Activity Report, p. 13.

2008/09, compared with around 33,000 tonnes between 1996/97 and 1998/99[274] . This drop in meal supply is linked to the fall in the quantity of crushed seeds. Studying the supply of oil and cake enables us to analyse their demand on the market.

b) Analysis of market demand for cottonseed oil and cake

According to the feasibility study for the autonomisation of the Direction Huilerie Savonnerie (DHS), world demand for oil[275] increased towards the end of the 1990s. Crushing activity has also risen sharply, but cotton production has fallen due to a decline in seed production. Cottonseed is also disadvantaged by its low fatty acid content (average yield of 18%), unlike its high-content competitors. Africa has a large oil deficit, and relies on imports to make up the shortfall[276] . The market for oilcake is also expected to grow during the 1998/99 marketing year. These trends are supported by the terms of reference of this study in the projected growth in demand for oil and cake in Chad and neighbouring countries[277] . It is therefore important to assess the need for cottonseed oil internally.

c) Demand for cottonseed oil in Chad

It is estimated that millions of litres of oil are consumed every day and every year. However, commercial outlets for the oil are not limited to Chad alone, but extend to two neighbouring countries: Cameroon and the Central African Republic. Export sales of co-products are not made directly by COTONTCHAD or the Direction Huilerie Savonnerie (DHS), but rather by Chadian customers of the latter[278] , particularly those in Moundou. Demand for oil is guided by the following characteristics: the reputation of the oils; purchasing habits; usage habits; the qualities sought and the image of the main products[279] . These characteristics were determined during group meetings with housewives organised in N'Djaména, Moundou, Doba, Koumra, Sarh and Abé- ché by ICEA during its diagnosis of the commercial strategy of the Direction de l'Huilerie Savonnerie (DHS) in 1991.

In terms of reputation, cottonseed oil is well known by households, which give each type of oil a specific name. Refined oil packaged in one-litre bottles is called "first quality", which is much sought after by households, and neutral oil packaged in redder, more fragrant drums is called "second quality", which is not much appreciated because of its negative consequences. In addition, endogenously produced groundnut, shea and sesame oil is called "artisanal", contributing to food self-sufficiency and competing on the market with cottonseed oil[280] . However, oil is purchased by housewives on a retail basis, in quantities rarely exceeding 1/3 litre and estimated at 250 ml on average per purchase. Based on the average price of a litre of oil on the market, this corresponds to a daily expenditure of around 150 FCFA per household, based on the weighted price for the 2007/08 season. Usage patterns vary according to the specific dishes prepared by housewives[281] . As a result, the qualities of oil sought are those measured by clarity, lightness, favourable taste, absence of odour, and absence of foam when cooked. Cottonseed oil and competing oils have the following characteristics:

- Neutral cottonseed oil: its strengths are its affordable price, its colour and its drum packaging, which makes it easy to transport and deliver to remote areas. But it has a bad

[274] E. Mbainaissem, 2013, p.103.
[275] P. Texier, 1994, "Huile de coton en Afrique, une industrie récente, coton et Développement", pp. 1117.
[276] *Ibid.*
[277] *Ibid.*
[278] Direction Huilerie Savonnerie, 2016-2017, Rapport d'activités commerciales Moundou (Tchad), p. 7.
[279] *Ibid.*
[280] F. Padacke, 2010, pp. 78-83.
[281] *Ibid.*

odour, and its use is confined to fritters and deep-fried foods, for which it provides good colouring, and because of its slow evaporation rate;

- Refined cottonseed oil: its strengths are its colour, smell and reputation. But it's a seasonal product with a high price tag, and its packaging in one-litre plastic bottles in a cardboard box is unsuitable;
- Peanut oil: its strengths lie in its good flavour and bright, yellow colour. It is used to prepare sauces, but does not give enough colour to fritters. According to DHS officials[282] , small-scale production does not allow the oil to be dried, leading to a rapid deterioration in quality.

d) Players in the oil and cake markets

Cottonseed oil is sold to wholesalers, retailers and consumers (households). Meal customers also include wholesalers, livestock farmers and agri-breeders. These wholesalers are distributed geographically in the following towns: Moundou, N'Djaména, Sarh, Koumra, Doba and other towns in Chad[283] . There are fewer wholesalers in N'Djaména than in Moundou, which tends to account for the bulk of sales. Export sales are mainly made by traders in Moundou, who take advantage of their proximity to the CAR[284] . A large number of customers are not traders, but private individuals from Moundou or elsewhere, who obtain larger quantities than an individual needs. The latter have not paid the same taxes as the traders, which represents unfair competition. As a result, customers are discouraged by the obligation to buy DHS products from influential non-traders who have easier access to purchasing authorisations[285] . These authorisations, although obtained in the name of the traders, are often sold back to them at a substantial margin.

However, oil deliveries are subject to a quota allocated to customers by oil mill managers. The allocation of quotas for DHS finished products is highly sensitive, as the principle is not always rationally applied and requires a high level of involvement from DHS managers[286] . Wholesalers' quotas were allocated in accordance with Decision No. 002/2008/DG/DHS on the marketing of DHS finished products. In accordance with this decision, a commission was set up within COTONTCHAD-SN by Decision No. 001/DAMG/DG/2008 of 22 October 2008[287] to examine applications for approval of DHS traders. Decision No. 002 stipulates that the approval of customers is the sole responsibility of the General Management of COTONTCHAD-SN after examination of the files by the commission[288] . Quarterly quotas will be allocated and any quota not collected within a period not exceeding one week will be allocated to other approved customers. Goods must be collected in the following stages:

- Prepayment by the wholesaler to the oil mill's bank account of the amount of the purchase authorisation granted by the DHS in the past and now by the General Management of COTONTCHAD-NS;
- Invoicing by the DG after receipt of the credit advice from the bank or the certified cheque;
- Lastly, the product is collected by the wholesaler once authorisation has been obtained

282 Direction Huilerie Savonnerie, 2015, Rapport d'activités commerciales, Moundou (Chad), p. 17.
283 Direction Huilerie Savonnerie, 2012, Rapport d'activités de la production et de la commercialisation, p. 25.
284 *Ibid.*
285 *Ibid.* p. 27.
286 Tableau de Bord de la COTONTCHAD-SN sur les activités de la production et de la commercialisation, pp. 19-23.
287 *Ibid.*
288 F. Padakc, 2010, p. 98.

from the DG[289] . It is not known what consumers' tolerance threshold is for oil, but it seems that the price elasticity of sales may yet remain positive, as demonstrated by the commercial policy of wholesalers at the start of the rainy season. They are stockpiling large volumes of oil from March and April onwards. Withdrawal takes place at the height of the rainy season, enabling wholesalers to almost double their prices. Oil prices at DHS level are set by a commission made up of some of COTONTCHAD-SN's key departments[290] ; a price scale drawn up on the basis of the commission's proposals is published periodically.

In short, cotton marketing requires organisation between the players to ensure that the market runs smoothly. This must be based on agreements to open up the market after the cotton has been harvested by the producers. Cotton marketing follows almost the same process as production, depending on the systems for storage, sorting, weighing and payment. This area takes into account debt collection in the village associations concerned at COTONT- CHAD-SN. The prices at which seed cotton is bought from producers are the same in all of Chad's cotton-growing zones over the course of a crop year. The evolution of these prices varies according to the crop year and the qualities of seed cotton after COTONTCHAD-SN's evaluation. There are three qualities of seed cotton: first choice, second choice and third choice. The first choice has a high price, the second a medium price and the third a low price.

However, COTONTCHAD-SN collects seed cotton from producers and transports it to the ginning plant for processing into a finished product. It then exports the fibre to world markets, while the by-products are sold on domestic markets for household consumption, animal feed and soil fertilisation.

[289] E. Mbainaissem, 2013, p. 142.

[290] *Ibid.*

CHAPTER IV

THE SOCIO-ECONOMIC IMPACT OF COTTON PRODUCTION AND MARKETING IN THE MANDOUL REGION ORIENTAL

This chapter looks at the strengths and weaknesses of cotton production and marketing in the Mandoul Oriental region. However, the introduction of cotton growing in Chad in general, and in this region in particular, was successfully expanded by colonial administrators at the economic, social and political levels. Cotton production has occupied a large part of southern Chad, on which more than three million people depend directly or indirectly for their income. This is what makes "Chad's useful south". Before the oil era, when nine cotton ginning factories were set up, cotton accounted for a significant proportion of export earnings in this area. As a result, the sector plays an undeniable role in the socio-economic and political development of Chad, particularly in the Mandoul Oriental region, which is the subject of this chapter.

I - THE POSITIVE IMPACT OF COTTON PRODUCTION IN THE EASTERN MANDOUL REGION

As noted above, cotton production in Chad has been a major factor in socio-economic and political development at both national and local level.

A - Social impact

The social impact of cotton growing enabled us to analyse the local community environment, i.e. the improvement in the living conditions of the rural population from a Western perspective.

1) Improving living conditions for the population of Mandoul Oriental

By studying the evolutionary trajectory of farmers, it is possible to classify them into more or less homogeneous sub-groups[291] . A farmer's membership of a sub-group has a certain stability over time. Given the objective of the study, the typology proposed is not exhaustive, but it does help to understand how the farmers in question have evolved over time to become what they are at the time of the surveys.

Two relatively simple and easily observable criteria were favoured: the main crop on the one hand and technical specialisation on the other. Since the colonial period, cotton growing has gradually exacerbated the differentiation between farmers. The principles of communal harmony and solidarity that levelled out differences from the bottom up are disappearing. Over time, the forced or coercive nature of cotton growing is increasingly forgotten, giving way to its voluntary adoption. This increases the technical and economic importance of cotton in the individual decisions of farm managers and the new agrarian landscape system. In the traditional village of the Man- doul Oriental region, the huts are often

[291] J. Chappelle, 1980, *Les peuples tchadiens, ses racines et sa vie quotidienne,* Paris, L'Harmattan, p. 102.

The roof is made of straw, the walls of rammed earth, and the inside is plastered with clay mixed with cow dung"[292] . These days, however, it's the "rectangular" shape that predominates, with roofs increasingly made of corrugated iron sheets, walls increasingly of cement bricks, the interior floor plastered with cement, and walls sometimes painted with industrial paint[293] .

In the eyes of the peasant, the sheet-metal house represents an outward sign of wealth. Farmers prefer brick and tin houses for their durability and solidity, their comfort and the social prestige they bring. The number of brick and tin houses increases with the importance of cotton cultivation in the system. It can therefore be argued that cotton growing makes a remarkable contribution to improving rural housing. Although the income from cotton accrues to the man, the woman benefits from the effects of the modernisation of the cotton production system, even though her work on the farm has become relatively more important than that of the man, particularly in the presence of mechanisation. Three facts stand out quite clearly:

The first, and most important, concerns reducing the "arduousness of women's work, both in the field and in the household"[294] . We compare women's current situation with that which prevailed when cotton growing was still marginal in production systems, as was the case in 1960. Ploughing tends to be done by men, and weeding by women has become less arduous. Ploughing reduces "weed cover, but the soil is also less hard when weeding, which makes the work relatively quicker and less tiring"[295] .

The use of herbicides reduces the time and effort required for manual weeding of cotton plots and especially cereal plots (rice, maize, millet, sorghum), where young plants sometimes resemble certain undesirable grasses and therefore need to be weeded.

One disadvantage that often comes to light is that harvesting work, which is mainly carried out by women, is taking longer and longer as a result of the increasing size of the plots in cotton-based systems. Housework is also becoming less arduous. Cereals are the main foodstuffs eaten by farmers, even those who prefer to grow yams. The most popular cereal-based dishes involve husking millet and sometimes grinding the grains into flour[296] .

For a long time, hulling and grinding were carried out manually by women, using local equipment (mortar, pestle and crushing stone). These operations required a great deal of time, physical effort and know-how. However, since the introduction of cotton growing, they are increasingly being carried out by semi-artisanal hulling machines, in return for relatively easy payment following the increase in income. In addition, women are now responsible for drawing water. This was done mainly at the water points closest to the village, from 100 m to 500 m or more depending on the case. Carrying the water to the village, in containers often made of "clay", on tracks with sometimes slippery slopes, was still not an easy task for the women. The daily chore of fetching water has become less and less arduous with the advent of boreholes in villages"[297] . Although these wells have sometimes been dug as part of government-funded programmes, their upkeep relies on the awareness and funding of local people. In the Mandoul region, more than 150 boreholes are operational thanks to income from cotton[298] . Firewood was mainly transported by women. This work is increasingly

[292] Interview with Yengar André, Koumra, 20 June 2017
[293] Interview with Yengar André, Koumra, 20 June 2017
[294] Interview with Mounmotoi Angeline, Bessada, 18 June 2017
[295] Interview with Ngar-idjimte, Bessada, 18 June 2017
[296] Interview with Mounmotoi Angeline, Bessada, 18 June 2017
[297] Interview with Ramadi Céline, Koumra, 21 June 2017
[298] B. Verardo et al. 2004, "Analyse de l'impact social et de la pauvreté Réforme du secteur coton au Tchad- Analyse qualitative ex-ante- Première phase", pp. 47-49.

carried out by cattle cart, over distances that have become longer as a result of the growth in demand due to demographic growth. The cart allows women to build up larger stocks of firewood.

The second fact concerns the time saved by rural women, thanks in particular to mechanisation[299] compared with women in manual farming systems. It is true that the overall volume of women's work has increased in the presence of mechanisation compared with that of men, but compared with the situation of women in non-mechanised systems, certain operations are carried out more quickly and more easily as a result of improved means of transport (bicycle, motorbike, cart) and increased income.

The time saved enables some women to take part in information and training sessions on self-determination; others are members of profit-making associations[300] such as shea soap producers and market gardeners like those in the village of Koko, 15 kilometres west of Koumra. Other women farmers take part in training and information sessions on sexually transmitted diseases, including HIV/AIDS.

The third fact concerns the increase in the income of women farmers. Although the income from cotton growing is managed by the man, the wives interviewed acknowledge that they receive a substantial part of it, even if this is sometimes at their insistence. Some wives have managed to get their husbands to entrust them with the income from the cotton, and they play the role of safekeeper[301] . Furthermore, the practice of growing groundnuts before cotton or in the normal rotation has led to an increase in the area under groundnuts. This has led to an increase in groundnut production and a rise in women's income, since they are the ones who manage the crop.

However, without having explored in depth the question of the place of women farmers on the farm, we can see that they are becoming increasingly important in the quest for emancipation and self-determination. So, even if cotton growing has led to women working more than men on the farm, it has to be admitted that, in the villages studied, the situation of women has improved relatively compared with the 1960s.

2) The creation of schools and health centres

a) The creation of schools

The colonial administration had introduced Western education into the cotton-growing areas in order to train agents in the operation of the seed cotton markets and the farming system[302] . This is why there are schools in the major towns of Chad and in the provinces. In Koumra (capital of Mandoul Oriental), we counted nine (09) primary school centres, including three (03) Islamic education centres[303] .

These schools were set up following the introduction of cotton growing in the region, and now operate thanks to contributions from cotton growers and aid from the Chadian government. According to parents in the villages, the cotton profits enabled us to enrol children in school and pay for their school supplies during the school year. The presidents of the APEs (Parents' Associations) ensured that parents paid their contributions when the cotton was paid, so that the quota for each child enrolled in school could be taken off[304] .

[299] Geneva Round Table IV, 1999, Sectoral meeting on rural development. Diagnosis and strategies. Ministry of Agriculture, p. 65.

[300] Interview with Madjissembaye, Koko, 24 June 2017

[301] Interview with Madjissembaye, Koko, 24 June 2017

[302] D.E. Gardinier, 1986, "Enseignement colonial français au Tchad (1900-1960), *Afrique et l'Asie Modernes*, pp. 59-72.

[303] Madana Nomay, 2001, *Les politiques éducatives au Tchad,* Paris, L'Harmattan, p. 196.

[304] Interview with Rassemadji, Begue, 24 June 2016

This contribution from the parents enables the parents' presidents to pay for "equipment and ensure the salaries of the teachers"[305] . Cotton money is the only income that allows us to write 6 to 8 children without regret, because we can't sell the millet reserve to ensure that. What's more, the purchase price of millet is not in the parents' favour, as is the case for cotton. In fact, the rebates granted by COTONTCHAD have contributed to the construction of classrooms and boreholes for pupils[306] .

b) The creation of health centres

The creation of medical centres in the Mandoul Oriental region is a major step forward in rural development, as people need health care to be able to carry out their activities properly. Farmers have told us that there's nothing better than good health. The presence of medical centres in the villages limits the distance travelled by farmers to hospitals in towns, such as the Goundi hospital, the regional hospital and the Seymour hospital in Koumra[307] . These hospitals liaise with the medical centres to limit the number of cases of illnesses that lead to death. Patients are transferred to these hospitals when their equipment is limited or when health workers are unable to carry out operations, or when they are suffering from cancer or high blood pressure[308] . Cotton growers are therefore better off in medical centres to deal with the illness that prevents them from combining their efforts in the fields and in many other professional activities.

3) Infrastructure development

Cotton growing not only benefited the colonial administration, which set up ginning factories to consolidate its development, but also enabled the local people to transform their environment from its natural state. Cotton revenues contributed to the construction of the Doba-Koumra, Sarh and Péni-Bédjondo roads, and secondary roads were asphalted[309] . The roads were built by the colonial administration and the local populations, enabling seed cotton to be evacuated and inputs to be placed in the A.V. In addition, we noted the creation of new companies such as the S.T.G (Société Générale Tchadienne) in Kou- mra, the capital of Mandoul Oriental, and the modern market was created in 2013[310] . In Mandoul, there are women's groups involved in rural development under the leadership of PRODEL (Projet de Développement Local), BELACD (Bureau d'Étude et de Liaison des Actions Caritatives du Diocèse) and ONDR (Office National de Développement Rural). These organisations have contributed to the well-being of the local population through the construction of shops, sewing workshops, agricultural equipment for the collective fields of rural women and the construction of meeting rooms in the townships. These infrastructures can be found in Koumra and in the sub-sectors of the region. It should also be noted that the

Cotton revenue contributes to the construction of administrative services: the prefecture, the sub-prefecture, the town hall, the governorate, the labour inspectorate and the Koumra courthouse.

B - Economic impact

The economic impact of cotton growing extends nationally and regionally through its spin-offs. This being the case, cotton is a vector for development, especially in rural areas, due to

305 Interview with Rassemadji, Begue, 24 June 2016
306 Interview with Sadjinan, Koko, 24 June 2016
307 Interview with Mbaitoloum, Koumra, 12 July 2017
308 Interview with Madjadoum, Koumra, 10 June 2017
309 Report on the action and development plan in Chad, p. 13.
310 Interview with Madjeyengar, Koumra, 12 July 2017

access to agricultural equipment and cultivation techniques.

1) The economic impact of cotton growing at national level

Cotton is by far the most important export crop, accounting for almost 80% of total exports[311]. In the 1990s, cotton exports brought in on average twice as much as livestock exports[312]. As a result, cotton remains the only source of monetary income for those who grow it, beyond any uncertainties that may arise, unlike food crops whose marketing is more uncertain.

The cotton industry has a very significant indirect impact on whole swathes of the country's economy. It really supports the few companies in the formal sector. Banks and insurance companies, as well as many transport companies throughout the country, depend to a large extent on the spin-offs of cotton production. Since then, the cotton sector has enabled cotton growers to buy goods from industrial companies in southern Chad, in particular the breweries in Logone Occidental (Moundou): Manufacture de Cigarettes du Tchad (M.C.T) and CYCLO TCHAD[313].

In addition, there are many other companies in Chad promoting its development. However, this sector is still underdeveloped and accounts for 9% of GDP. Indeed, the COTONTCHAD-SN textile industry, with its 9 cotton ginning plants and the S.T.T (Société Textile du Tchad), includes a fully integrated spinning, weaving and handling complex at Fort-Archambault (Sarh)[314]. There are also a number of small businesses, including two refrigerated slaughterhouses in an unused meat and leather industry complex, three cottonseed and groundnut oil mills in Moundou, and three rice mills and two dairies in Fort-Lamy (N'Djaména)[315]. In addition, there are industries covering the energy sector, construction, mechanical engineering, printing and a clothing factory in Fort-Lamy (N'Djaména)[316]. Of these, all but cotton and meat are exported, but the remainder is consumed on the domestic market. Cotton has been used to develop food crops for the family.

2) Cotton growing and the development of food crops in the Mandoul Oriental region

a) Cotton growing

Numerous studies have been carried out on the development of the cotton-growing areas of West and Central Africa. The development of cotton production is based mainly on the following factors:

- Financing, credit, guaranteed prices ;
- Training, research, support and extension.

The supply of inputs (selected seeds, fertilisers, insecticides and herbicides) and means of production (agricultural equipment at the right time and on credit[317]), as well as a guaranteed price and training for the farmer, were reinforced by the determination of the C.F.D.T. and I.R.C.T.. This system existed for a very long period, from the 1950s, in order to develop cotton production.

---

[311] World Bank, 2001, *Global Development Finance. African Economic Outlook* OECD/BAFD, p. 19.

[312] B. Donon, 1998, *Généralité ouvrage de synthèse de l'Afrique noire*, Chad, p. 205

[313] CYCLOTCHAD is a French company that set up a cycle assembly workshop in Moundou in 1957. Its production capacity is 10,000 cycles a year, Moundou (Chad), p. 32.

[314] Fort-Archambolt: This name refers to the former town of Sarh.

[315] From colonial times to the early 1970s, Fort-Lamy was the name given to the present-day city of Ndjamena, a city named after the cultural revolution launched by Ngarta Tombalbaye.

[316] Commission of the European Communities and Direction des échanges commerciaux et du développement, vol.10, République du Tchad, p. 27.

[317] Y. Bigot and G. Raymond, 1991, " Traction animale et motorisation en zone cotonnière d'Afrique de l'Ouest : Burkina Faso, Côte-d'Ivoire, Mali ", CIRAD-DSA, CIRAD-IRCT, *Collection Documents Systèmes Agraires*, n° 14, Cirad, p. 95.

However, from 1965 to 1985, the price of cotton on the world market was very buoyant. Certain factors favoured farmers, enabling them to equip themselves with animal traction and to modernise by learning techniques favourable to cotton and food production[318] . To achieve this, the State intervenes by subsidising agricultural equipment for farmers. Farmers have access to animal traction, mechanised machinery, ploughs and carts on credit from the ONDR[319] . NPKSB fertiliser, urea, pesticides, batteries and sprayers are supplied by COTONTCHAD-SN to help the young cotton plants grow well and produce good yields. This is confirmed by our following informant:

Many farmers prefer to hire the tractor and animal traction for one-day ploughing instead of 5 days of weeding per group of people. Animal traction was introduced in Mandoul in 1956 by the ONDR. Prices for harnessed and manual ploughing vary in rural and urban areas. However, once they had received their cotton money, the farmers organised themselves for the following year's crop. They bought groundnut seed, peas and cereals to survive. The cotton money enabled them to trade, develop livestock and acquire the material goods they needed[320] .

In Chad, particularly in the Mandoul Oriental region, farmers who have farming equipment compete for land in order to increase yields each season. Added to this are the "efforts of women farmers who own pairs of oxen, carts and build tin houses"[321] . Some women trade in cereals and traditional beers (bili-bili, djala and argui). In the course of our research, informants told us that after the cotton payment, "it's a feast in the village" because people are so happy. Even those who didn't grow cotton benefited because the money circulated. The big traders in the villages and towns of Mandoul Oriental are farmers, who invest their money in trade, buying land and giving credit to civil servants. The latter repay with interest at the end of each month. They also save money in banks. Cotton growing is therefore a lever for development in both rural and urban areas. From then on, the cotton cultivation technique recommended by the Agents Cotonniers de Terrain (ACT) in the agricultural calendar can be used for food crops. This has enabled farmers to improve the sowing system and crop rotation.

b) Cotton and the development of food crops

Since the mid-1960s, scientists have sometimes questioned the transformation of the rural environment in these areas, without indulging in complacency about cotton growing. But everyone agrees on several significant facts about the place of cotton in the production system. Food crops have gradually been integrated into the cotton sector to improve cereal production in particular.

In the Mandoul Oriental region, cotton production has boosted food crop yields through the after-effects of organic fertilisers applied to the cotton plant. The food crops grown in this region are millet, beans, maize, groundnuts, rice, potato peas, sesame and cassava. Supervision of the farmers has enabled them to grow the following rotation crops: "cotton-peanut-cotton, cotton-maize-sorghum, maize-bean-sesame"[322] . This shows that farmers do not grow the same crop twice on the same farmland. Naturally, the soil contains mineral substances, but the use of fertilisers enhances its fertility for the growth of the crops grown there. In Man- doul, "groundnuts, millet, maize, sesame and cassava are the dominant food

[318] M. Braud, 1990, *La filière coton en Afrique de l'Ouest et du Centre, un îlot de progrès dans un Océan de morosité*, Paris [FR]: CIRAD-IRCT, p. 123.

[319] *Ibid.* p. 127.

[320] Interview with Allarayem, Ngandou, 22 June 2017

[321] Interview with Hoiadi Sidonie, Kotkouli, 16 June 2017

[322] Interview with Djasngombaye, Kemkian, 23 June 2017

crops after cotton. Some areas of the region are not suitable for growing maize and cowpeas"[323] . Cassava is grown by farmers on all types of soil, and is the most popular crop when grown in woodland, as it gives good yields even under trees. Cassava roots are very rich in soil fertiliser and alternate with either millet, groundnuts or sesame. Maize, on the other hand, "requires organic manure or animal manure in most cases"[324] . Food crops are therefore grown for household consumption, with some for sale to meet other needs.
In addition, the practice of growing woody crops is recurrent in the Mandoul Oriental region, especially during the rainy season, as almost all the farmers are interested in these crops. Those living in the riverside villages grew vegetables, maize, potatoes, taro, tomatoes, cucumbers and aubergines during the dry season, which brought in a lot of revenue for their projects. It should also be noted that farmers in Mandoul are interested in growing pima because of its very high price on the local market. Food crops have thus enabled the development of livestock farming, which is the lifeblood of Chad's economy. However, crop rotation in an agricultural plot requires the soil to be left to rest, as recommended by the cultivation technique.

c) The system from fallow to soil fertilisation

The most common traditional land-use system consists of a 5-10 year cropping phase, followed by abandonment of cultivation to allow fertility to develop, or weeds or parasites to invade[325] . In many cotton-growing areas of southern Chad, this principle of soil fertility management has evolved under the combined effects of the factors mentioned above. One of the major changes observed in eastern Mandoul is the considerable reduction in the length of fallow periods. As well as renewing cropping conditions by restoring fertility, fallow land performs multiple functions: erosion control, land management, biodiversity conservation, fodder production, wood of all kinds and medicinal plants. Studies into the dynamics of fallow land[326] show that there are not many alternatives to natural fallow. The main suggestions for the sustainable management of cultivated soils are to use fertilisers combined with biological methods, given the dominant role played by organic matter in the fertility of these soils. Although some areas have been subjected to major demographic pressures on the availability of natural resources, fallowing remains the main practice for restoring fertility. However, we have mentioned two types of fallow duration: long-duration fallow, due to the availability of land and the low rate of farmers, and short-duration fallow, due to the lack of availability of land[327] . Generally speaking, short-duration fallow lasts between 2 and 3 years, which does not favour certain agricultural plots. Farmers implement practices to improve soil fertility. Soil fertility is the ability of the environment to sustainably meet the needs of rural populations through the production and management systems they implement. In so doing, fertility is the result of the integration of man and the environment, of an evolving social construct. Fertility depends on the physical and biological characteristics of the environment, and on the needs and resources of the social groups that use and shape it to create an

[323] Interview with Djasngombaye, Kemkian, 23 June 2017
[324] Interview with Bolnan, Kemkian, 23 June 2017
[325] C. Floret et al. 1993, *La jachère en Afrique tropicale*, Paris, Unesco, MAAB dossier, p. 16.
[326] C. Floret and R. Pontanier, 1999, " Raccourcissement du temps de jachère, biodiversité et développement durable en Afrique Centrale (Cameroun) et en Afrique de l'Ouest (Sénégal, Mali) " *In Agriculture tropicale et subtropicale, troisième programme STD, 1992-1995.* Final report, ORSTOM/IRAD/IER/ISRA, p. 245.
[327] M. Gaide, "Au Tchad, les transformations subies par l'agriculture traditionnelle sous l'influence de la culture cotonnière", L'agronomie tropicale, vol. XI, No. 5-6, p. 47.

environment that is favourable to achieving their objectives[328] . Maintaining soil fertility in the Sudanian and Guinean zones depends mainly on protecting the clay-humus complex. This involves the systematic restoration of organic matter to an acceptable level of 0.8 to 1.2%[329] . To this end, research and development projects and NGOs are interested in environmental issues. These include agroforestry, erosion control, improved fallow, the use of animal waste and crop residues, and more rational use of the mineral fertiliser provided by cotton growing. In addition, the use of animal manure to improve soil fertility is very important, through animal dung from the livestock pen and that returned directly to the plot during grazing. Some farmers have herds and practise rotational grazing on their fields. This technique is commonly used by the Fulani, enabling them to restore the fertility of their maize and millet fields in the immediate vicinity of their homes[330] .

Farmers are starting to use the system, but to a very limited extent. It is true that this practice is only rewarding for producers with a sufficient number of livestock, which is difficult for the majority of farmers. So, when opportunities arise, they sign manure contracts with transhumant or sedentary Fulani herders. The contracts are based on a simple principle, requiring no financial or material compensation. They are based on exchanges of profit: use of crop residues for the farmer; profit from animal dung for the farmer[331] . This is what happens in our study area, more specifically in the cantons of Koumra, Matekaga, Ngangara, Goundi and Dobo.

Contracts with livestock farmers are occasionally limited in time, which is not enough to effectively improve fertility. This process of using manure in these agro-pastoral areas is in its infancy. It is taking place against a backdrop of deteriorating agricultural production potential, with soil in particular still limited. To this end, the effectiveness of animal dung is a source of hope for farmers and has been scientifically proven, as Landais points out at[332] : "in addition to fertilising, manure plays an important role in soil structure, water retention capacity and stability through its organic matter...". Indeed, the concept of fallow goes hand in hand with rotational cropping.

However, cotton and food crops have enabled farmers to include domestic livestock and transhumant herders in their systems.

d) Livestock farming integrated with agriculture

Livestock farming accounts for between 15% and 20% of the country's economy, after cotton[333] . The dominant livestock farming system is extensive, essentially dependent on pastoral resources, the most important of which are grazing and the availability of water. However, the Sudanian zone, considered as the agricultural area par excellence, is also a sedentary livestock area. It is worth mentioning that, due to increasingly favourable ecological conditions, it is now an excellent host area for transhumant herders, who are tending to stay longer or even become sedentary. Population density and the influx of transhumant herders

[328] *Ibid.*
[329] *Ibid.* p. 248.
[330] P. D'Acquino, 1995, p. 132.
[331] C. Haessler et al. 2002, Développement du cheptel au Sud du Tchad: quelles politiques pour l'élevage des savanes? In Jamin J.Y., Seiny Boukar " *Savanes africaines : des espaces en mutation, des acteurs face à de nouveaux défis* ". Conference proceedings, May 2002, Garoua, Cameroon, N'Djamena, Chad, p. 19.
[332] E. Landais, 1993, Systèmes d'élevage et transfert de fertilité dans les zones de savanes africaines. Les systèmes de gestion de la fumure animale et leur insertion dans les relations élevage et agriculture. *In Cahiers Agricultures,* volume 2, number 1, pp. 9-26.
[333] Ministry of Livestock, 1998, Réflexion sur l'élevage au Tchad. Rapport principal, pp. 77-79.

are creating strong local pressures on the environment. Indeed, 85% of farmers raise cattle for farming purposes and small ruminants for economic reasons. During the lean season, cotton cake is sold to farmers to feed their livestock. For this reason, cotton cultivation plays an important role in the development of food crops and livestock farming[334] . Furthermore, Lhoste stated:

> As opposed to the extensive system, the system includes animals whose management is linked to the farm. This means that, for at least part of the year, the animals are housed in outbuildings on the family plot; it also means that part of the feed is distributed and food stocks are built up and managed. In general, the system also reflects a certain form of intensification and more individualised management of the animals[335] .

This definition is challenged by the current trend in livestock farming, which is integrated into the farm in Mandoul Oriental. For example, there are situations where livestock farming is a support activity, adding value to by-products and cleaning up the vegetation around the fields, as in the case of farmers with one or two pairs of oxen for ploughing. These farmers keep their animals away from the village in order to avoid theft of their oxen. However, there are other models where livestock farming has become an essential component of the agrarian system for both ecological and economic reasons[336] . In fact, three systems are classified:

> The transhumant system: this is the oldest and most widespread practice, and differs from the others in several respects: the extent of transhumance, its duration and the livelihood of the farmer, who depends on the animals[47] (food, income). The spatial and temporal movements of herds are linked to their geographical location and the availability of pastoral resources (fodder, water). Mobility is essential to the survival of the group; it is also a source of social value and is acknowledged as such by individuals whose qualities are recognised by all. Travel is carried out in anticipation of a certain level of constraint. The right choice of route is one that brings the animals to the start of a new phase in the pastoral calendar in the best possible conditions. The types of farming system in the study area should be illustrated.

> The semi-transhumant system: is defined as an intermediate system between transhumant livestock farming and livestock farming associated with agriculture. The system involves both an agricultural obligation that keeps producers in their territory and a pastoral constraint that leads them to carry out seasonal transhumance of short duration (2 to 3 months) and short range (20 to 30 km on average). In the Sudanian zone, mobile pastoral practices are tending to stabilise with increasing recourse to semi-transhumance, which seems to be a new form of adaptation to the context of saturated space.

> The sedentary or peasant system: this system integrates animal and agricultural production. Farming is the foundation of the system on which the producers' way of life is based. As the system develops, the animal becomes an integral part of the production system: animal traction, transport and manure to fertilise the fields. This is the most widespread farming system in the Sudanian zone, especially as it requires animal traction[337] .

However, the southern region is an area where livestock breeders settle down and stay during hot periods. All types of livestock farming systems coexist here: extensive systems based on transhumance, agricultural systems, sedentary and semi-sedentary systems used by local

(Senegal). INAPG thesis, p. 312.

46 J. C. Clanet, 1982, "L'insertion des aires pastorales dans les zones sédentaires du Tchad central",

334 C. Haessler et al. 2002, pp. 7-11.

335 P. Lhoste, 1987, L'association agriculture-élevage : évolution du système agro-pastoral au siné-saloum.

47 Ibid.

Cahiers d'Outre-Mer, vol. 35, no. 139, p. 63.

337 E.Vall et al. 2000, Études des pratiques et stratégies paysannes de traction animale dans les zones des savanes cotonnières du Cameroun, Tchad et RCA, Prasac/Irad, pp. 21-24.

populations and the Mbororo. As a result, it is difficult to obtain real figures for cattle and small ruminants. Among sedentary livestock farmers, the Ministry of Livestock[338] , in the absence of a census, bases its figures on vaccinated animals and is therefore subject to significant bias. Nomadic livestock farmers distrust the government services and avoid having their animals vaccinated regularly; it is impossible to know the exact number of animals kept by a farmer through vaccination. This means that transhumant herds are rarely taken into account by the Ministry of Livestock, even if estimates have to be made.

In the Mandoul Oriental region, the Ministry of Livestock estimated that there were 27,000 head of cattle in 1997 and 39,000 head in 2000. This figure rose to 54,000 head in 2007 and 85,000 head in 2015[339] . It should be stressed that in the current context of the region, apart from transhumant herders, there are no livestock farmers who are not involved in farming. Even so, few farmers do not keep a few head of cattle. Thus, for the purposes of this study, the groups considered to be farmers were those from the village communities (Sara, Gouleye, Nar, Ngambaye, Gor and Daï). On the other hand, the groups considered as breeders and also as non-natives with dominant activities such as breeding, whether they are sedentary or transhumant (Misseriers, Arabs and Mbororo). In fact, the term "agro-breeder" is used for sedentary people who combine farming with livestock rearing, and "transhumant breeders" for transient breeders. These activities play an important role in agricultural development. They show that farmers and breeders are complementary in terms of animal and soil management.

C- Environmental impact

The environmental impact is characterised by the colonisation of trees, the management of agricultural space and their role as land markers.

1) Tree colonisation and land management

In traditional farming systems, trees are a vital element because of the roles they play, particularly in protecting the soil, marking out land and providing fuel. The many functions performed by trees in the management of agrarian space demonstrate their justification and their place in conservation for man in his environment. The functions of trees have been the subject of numerous scientific publications in many disciplines, including the collective work published in 1980 by ORSTOM[340] , entitled "L'arbre en Afrique tropicale, la fonction et le signe", edited by Pelissier, which shows the role played by trees in agrarian dynamics in sub-Saharan Africa. For him, the evolution of African agriculture explains its sedentarisation and intensification, which did not involve the elimination of trees, but rather their integration into crops. To bear witness to the many functions performed by trees in the history of mankind and their environment[341] , this study looks at the evolutionary aspect of agrarian practices integrating trees in the management of space and fertility.

In the Sara region, for example, certain trees are used not only to fertilise the soil, but also to perform ritual functions. They also contribute to the food self-sufficiency of the local population and animals to a certain extent. Shea (kian) is consumed throughout the Mandoul Oriental region as soon as the first rains arrive, and its fruit is used to make household oil. Néré (mat) is consumed most during the dry season, and its flour is used to make porridge, while its seeds are used to make traditional maggie

338 Ministry of Livestock, 1998, p. 96.

339 Report on animal vaccination 2015 by the Ministry of Livestock in the Mandoul Oriental region, p. 7.

340 P. Pélissier, 1980, "L'arbre dans les paysages agraires de l'Afrique noire. In l'arbre en Afrique tropicale, la fonction et le signe", *Cahiers ORSTOM, Série sciences humaines*, vol. XVII, p. 4.

341 C. Seignobos, 1981, L'arbre et la cite dans la zone soudano-sahélienne 9exemple du Tchad et Cameroun) Revue de géographie du Cameroun, numéro 1, 1981, pp. 52-53.

(ndii). Néré (mat) leaves are used by traditional chiefs to venerate the ancestors in the village in order to obtain blessings and resist the evils that threaten the community[342] .

We note the presence of other useful trees: mango, guava, lemon, doum palm, roan and acacia, which are planted by the local population. However, fruit and non-fruit trees are preserved by the farmers in the farmland, even though some are eliminated. From then on, useful trees are popularised by development projects as part of land programmes and the fight against desertification. Trees play a major role in the region's traditional land tenure system.

2) Trees as landmarks

The contribution of trees to the assertion of land rights is well known to Africanist geographers:

The plant landscape is the visible imprint of the inherently inalienable land rights held by the first clearers and their descendants. While in natural law, land is only a matter of exploitation rights, the concept of ownership applies in its entirety to trees[343] .

Recent changes in the traditional land tenure system have further strengthened the role of trees in this area. Trees are becoming a marker of land ownership, and it is very common in regions facing land problems for trees to be a fundamental element in land appropriation strategies[344] .

During our survey, we realised that transhumant herders were not familiar with the system, and that if they were, they would have occupied large areas that constituted parks for their livestock. Agro-pastoralists have mastered the land tenure system by planting trees, which have become agrarian and habitable spaces. As a result, transhumance is known to be short-lived in search of pasture. Thus, the popularisation of environmental practices consists of planting trees in quantity in order to combat drought and desertification as well as erosion, which are factors in soil degradation.

II - THE NEGATIVE IMPACT OF COTTON IN THE EASTERN MANDUL

Cotton growing contributes to improving the standard of living of the local population and to national development. But it also represents a danger for the social stratum.

A - Social impact

The negative social impact of cotton production in Chad in general and in the Mandoul Oriental region in particular can be summed up in educational and structural terms.

1) Education

The introduction of cotton growing has enabled the majority of farmers to provide for their children's education. It's not a bad crop as such, but the fact that a large proportion of this income is used to enrol children in school means that, once they're in, they pick up another habit, which is disrespect. It should be stressed that modern education is the main cause of acculturation in African societies, which have not encouraged the promotion of the local culture that makes up the history of Africa in general and Chad in particular[345] . In the Mandoul Oriental region, 70% of the rural exodus is attributed to young people attending school[346] , who create just as many problems for their parents: rape of girls, unwanted

[342] Interview with Sillenengar, Ngomana, 24 June 2017

[343] Pelissier, 1980, p. 9.

[344] Seignobos, 1980, p. 62.

[345] Mbaisso Adoum, 1990, *L'éducation au Tchad, bilan, problématique et perspective*, Paris, Karthala, p.201.

[346] *Ibid.*

pregnancies, theft and murder. Those who have taken the trouble to stay with their parents during the holidays become lazy; instead of going to work in the fields and preparing for the new school year, they spend their time drinking in pubs, playing cards or football. Some go back to work on the land, but grow only a few groundnuts and sesame for their own needs, and are always under the care of their parents. As a result, the education system is wiping out certain traditions in rural areas, as traditional education[347] loses much of its value:

The refusal to go to initiation in Sara country (yo-ndoh), the abandonment of or failure to respect ancestral cult celebrations at a given period, the disappearance of traditional songs and dances as part of education[348] , because the colonial administration has transformed all this to its advantage.

2) The structure of the social life of the people of East Mandul

Cotton growing has certainly brought about social change in the region, but it has not promoted rural development in its entirety. We have mentioned housing, means of transport, difficult access to drinking water in some villages, health care and education as criteria for the modernisation that cotton growing brought to Chad. Yet the standard of living of the Chadian and Mano Douani population remains precarious. Indeed, before the introduction of co-ton farming in the region, people still lived in round huts with thatched roofs, as today's peasants do[349] . In some Mandoul villages, even straw houses are hard to build.

It was the combined efforts of a few farmers who built the tin houses and shops, financed not by cotton money or state aid, but by the farmers themselves. Many of the old men who were involved in cotton production are homeless and exposed to the elements[350] .

This does not motivate the farmers to give themselves to cotton production, for them it is a loss-making job. Another major problem is the lack of access to medical care and drinking water. Farmers were unable to seek medical treatment because the cost of prescriptions was so high. While it is well known that human resources are a factor in the development of a country or a community, in the production system the health of the people is paramount to the quality of the work they do.

Health districts have been created, sometimes 30 to 40 km away in certain localities, and the means of transport pose enormous problems insofar as the farmers have no money to acquire these goods because COTONTCHAD-SN delays the collection of the cotton and payment[351] .

In villages, people have to pay for borehole water for drinking and many other activities. This system therefore contributes to economic imbalance.

B- Economic impact

The economic impact of cotton growing stems from the abandonment of food crops as a source of the household economy and consumption. It also refers to the occupation of agricultural land and the involuntary sale of economic assets through investment in this cash crop[352] . This is what qualifies cotton production, from the fragility to the food balance. The farmers lamented, "cotton growing is against our will[353] " because it brings us nothing. It requires large sums of money, and producers were obliged to sell their millet or animals to pay for labour. Cotton reduces the family economy and ultimately creates famine because it

---

[347] Hassan Hkayar Issa, 1976, *Le refus de l'école, Contribution à l'étude des problèmes de l'éducation chez musulmane du Ouaddai (Tchad),* Paris, C.N.R.S, p. 208.

[348] Interview with Allangombaye, Kemkian, 15 July 2017

[349] Interview with Ngar-adoumbe, Kemkian, 15 July 2017

[350] Interview with Nantoyim-kingar, Kotkouli, 16 June 2017

[351] Interview with Djasngomta, Koumra, 18 July 2017

[352] E. Mbetid-Bessane et al. 2003, p. 113.

[353] Interview with Tchadingar, Koumra, 19 June 2017

occupies the granaries without being evacuated by COTONTCHAD-SN. However, farmers know where to invest for their economic development when the cotton system changes, and they do so just to have access to NPKSB and urea fertilisers for the development of food crops. What is hard to understand, however, is that cotton is still the dominant crop in some villages, even though it exhausts the soil through mechanised traction-based crop rotation, which encourages erosion[354] .

However, traditional agricultural and pastoral systems for exploiting natural resources suffer the consequences of drought, but are also responsible for increasing vulnerability to the risk of desertification. Similarly, so-called "modern" systems based on cotton, groundnuts and maize are also vulnerable to desertification: poorly managed irrigation leads to salinisation and then sterilisation of the soil; monocultures can lead to greater susceptibility to erosion and soil degradation as a result of prolonged soil denudation[355] .

Mechanisms for controlling access to natural resources and management methods have generally been in place for centuries by traditional societies and have led to relative situations of equilibrium. They have subsequently been disrupted by historical and demographic factors, which can be explained by population growth. They are no longer sufficient to satisfy the needs of the population, especially in terms of food. Increased agricultural production has often been achieved at the cost of increased pressure on resources and space: increased cultivated areas, less fallow land, and therefore loss of fertility and greater susceptibility to degradation[356] . Added to this is deforestation for firewood and other uses. In the Sahelian zone of Chad itself, the main resource came from livestock farming; periods of drought led to overgrazing towards the south, resulting in severe degradation[357] . When rainy periods return, the pasture quickly recovers, but there may be a loss of biodiversity and a reduction in regrowth areas where degradation has reached a point of no return. This analysis therefore enables us to assess the environmental impact of cotton production in Chad in general and in the Mandoul Oriental region in particular.

C) The environmental impact of cotton growing

Cotton growing has not only had positive impacts. There have been many negative repercussions from cotton growing in the cotton-growing areas of Chad, contributing further to the fragility of the population. The intensification of cotton growing has negative consequences for the environment, contributing to climate change, the decline in biological diversity and the acceleration of desertification. The study of the environmental impact of cotton growing therefore focuses on soil degradation and biodiversity.

1) The impact of cotton growing on the soil

The impact of cotton growing on the soil is due to traditional cropping systems involving inappropriate clearing and burning. These cropping systems, which include mechanised systems on the one hand and ploughing on the other, are responsible for the decline in soil fertility[358] . In traditional cropping systems, the cycle begins with manual clearing of the

---

[354] D. Marambaye, 2002, "Évolution des conditions paysannes de production du coton au Sud du Tchad et ses conséquences sur les stratégies des paysans", Rapport de Maitrise, PRASAC, N'Djamena, p. 86.

[355] G. Raymond, 1991, gestion de la fertilité des sols et production cotonnière dans le Sud du Tchad, IRCT-CIRAD, session "*L'agriculture et la gestion des ressources renouvelables*", workshop B1, pp. 77-82.

[356] *Ibid.*

[357] E. Landais, 1993, p. 58.

[358] J. Arrivets and D. Rollin, 2002, *Questions de fertilité dans la zone soudanienne du Tchad : Proposition d'un travail de recherche de développement utilisant des systèmes avec SCV, mission report*, Montpellier, CIRAD, p. 61.

plots, which involves burning the plants that temporarily enrich the soil and leaving the burnt stumps in place.
However, intercropping is then carried out with reduced tillage. Soil degradation is limited and, after a few years of cultivation, will gradually lead to fallow land. In addition, the practice of burning is harmful and impoverishing because the plant matter that has fed on the soil does not return to it to be transformed into nutrients. To do this, the nitrogen volatilises and the ashes are dispersed by run-off water or the wind, so that the loss of nutrients through fire is equivalent. This also disrupts the restoration of soil organic recharge. In addition, fire superficially cooks the soil, altering its surface structure. The result is a reduction in porosity and an increase in run-off and erosion. In the Mandoul Oriental region, the administrative authorities have issued a statement banning land clearing, abusive tree felling and the use of bush fires by farmers, but the phenomena are continuing. The degradation of the soil due to climatic phenomena is at the same time encouraging the advance of the desert[359] .
Mechanised cropping systems also have their drawbacks, due to the highly-developed farming techniques used to extend the areas sown and achieve high yields. However, after radical clearing, which involves the total stripping of plots of land, annual crops with low yields are grown on the ploughed land. Cash crops, such as cotton, are integrated into the crop rotation cycle, and intensive techniques such as NPK fertilisers and plant protection treatments are used. To achieve this, the return on investment in terms of time and energy spent on developing the plots often involves continuous cropping with little or no fallow. So, not only do modern farming practices increase degradation, but they also encourage farmers to focus more on profitability than on the regressive evolution of the environment and farmland.
In addition, all land clearing causes an imbalance in the soil, but depending on the degree of imbalance and the cultivation system that follows, the effects can be even greater, the extreme cases being those of radical and brutal land clearing, followed by intensive cultivation that is sometimes unsuited to the local environment. In this situation, Roose[360] points out that "the humus-bearing horizons are stripped away, leaving only a crusted, compact, inert, almost sterile mineral mass". Total land clearing has medium- and long-term consequences, as it interrupts the soil's fertilisation cycle, to which plant matter and nutrients are not returned. Clearing also has harmful effects in the short term, due to the degradation of surface horizons that it causes, resulting in a decrease in the level of organic matter. This results in a reduction in nutrient stocks, which can lead to soil acidification and aluminium toxicity, through selective erosion resulting in a depletion of fine particles[361] . Mechanised land clearing is all the more devastating because it rips out the root network that provides the soil's structure and hence its stability, strips the humus-bearing horizons and pulverises the superficial horizons while compacting the lower horizons. What's more, they are accompanied by total stump removal, giving rise to vast holes, or wetlands, which are not filled in, even during levelling operations. Not only does complete stump removal eliminate the rotting of the stumps, which could have enriched the soil with organic matter, it also eliminates the shoots that normally develop on the stumps and which are favourable to regeneration when the land is set aside[362] .

---

[359] B. Verardo, et al, 2004, p. 68.
[360] E. Roose, 1985, "Impact du defrichessement sur la dégradation des sols tropicaux", *le machinisme agricole tropical,* issue 87, p. 27.
[361] E. Roose, 1985. p. 29.
[362] J. Peltre-Wurtz, 1984, "La charrue, le travail, l'arbre", *cahiers, Sciences Humaines series*, volume 20, number 3-4, pp. 63-64.

Mechanisation has two types of impact on soil degradation. Indirectly, it encourages an increase in the area cleared and sown, as it saves time when ploughing and sowing[363] . This is all the more detrimental as plots become longer and longer in the direction of the slope. This exposes new land to aggressive weather conditions. It also directly disrupts the soil balance, particularly in light soils that are easily leached and destructured. Clay soils that have been moistened are more resilient, especially if they are rich in humus.

As a result, mechanisation brings about negative structural changes in the soil, disturbances that go hand in hand with an increase in the structural instability index and make the soil more vulnerable to erosion. Thus, as a marginal cultivation practice, Boli and co-authors[364] mentioned the risks of rapid degradation:

Motorised ploughing, which removes the entire root structure of woody plants and especially annual grasses in the first year, and exerts considerable pressure on the soil, produces a deterioration in two years that has been achieved over several years in plots where manual *ploughing* has been used.

At the same time, however, deep ploughing is harmful because it breaks up the soil to a depth of 20 to 25 cm. This accelerates leaching and erosion, which can sterilise the soil after just two or three cropping seasons, especially as erosion is increased by 0.5% slope[365] . In addition, it exposes deep materials to capping and then reduces the cohesion and strength of the soil. Finally, it dilutes the organic matter and, above all, buries it in clumps in the deeper horizons where the anaerobic conditions are unfavourable to its development[366] .

2) The impact of cotton growing on wildlife and animals

Cotton growing has a negative impact on wildlife and animals as a result of the use of plant protection products. During the course of the research, our respondents felt that cotton growing has an impact on ecosystems. They thought that the heavy pressure on the plant cover and the use of plant protection products due to this crop seriously threaten the sustainability of the "natural wildlife resources of protected areas. The most affected are reptiles, monkeys, rats and birds"[367] .

However, the classified forests of the southern zone, particularly the Mandoul Oriental, are the object of uncontrolled felling and clearing by local people in search of new land suitable for growing cotton. Indeed, the cereal grains left on the fields attract birds most of the time in the area, which used to have a permanent presence of avian species: "The woodpeckers that used to accompany our herds home are no longer there," Yemingaye tells us, since cotton is not eaten by these birds, compared with cereals[368] . Moreover, cotton growing has harmful effects on domestic animals because it :

It reduces grazing areas by invading the land and causes livestock losses. Animals die of intoxication or poisoning after eating feed that has come into contact with cotton treatment products. The animals most affected are cattle, sheep, goats and poultry[369] .

We can also note the negative impact of cotton production on insects in the study area. The overuse of pesticides is a major threat to the survival of the insects that colonise the fields

---

[363] E. Roose, 1985, p. 31.

[364] Z. Boli et al, 1993, Effet des techniques culturales sur les ruissellements, l'érosion et la production du coton et maïs sur un sol ferrugineux tropical sableux. "*Recherche de systèmes de culture intensifs et durables en région soudanienne Nord-Cameroun, cahier ORSTAM, série pédologie*, vol. 28, no. 2, p. 54.

[365] G. Charriere, 1984, "La culture attelée : un progrès dangereux", *Cahier ORSTOM, Sciences Humaines series*, volume 20, number 3-4, p. 85.

[366] *Ibid.*

[367] Interview with Maikoe, Ngandou, 22 June 2017

[368] Interview with Yemingaye, Ngandou, 22 June 2017

[369] Interview with Yemingaye, Ngandou, 22 June 2017

(bees, termites, ants, locusts, whiteflies, caterpillars, butterflies, bacteria and champions)[370]. The bacteria, termites and earthworms responsible for decomposing plant debris and turning over the soil's surface horizons are killed off. As a result, biodiversity is thrown out of balance, leading to a blockage or slowdown in the renewal of soil humus[371].

The local people believe that the decline in the quality and, above all, the quantity of honey in their area is undoubtedly due to the use of pesticides. That's why Allarayem said: "In recent years, we can't get honey from the forests. Before, we could harvest at least twice (02) a year. But now it is often difficult to fill a rock in a year"[372]. This is considered to be destruction of nature, which does not leave man indifferent to his environment.

3) The impact of cotton growing on humans

Man's combined efforts to achieve economic development have adverse effects on his survival. The survey revealed accidents linked to the use of chemical pesticides. Skin and eye diseases, dizziness and even blood pressure were all reported. Numerous cases of death have been recorded in the Mandoul Oriental region, linked to poisoning by cotton chemicals. We found that these accidents were mainly due to poor practices:

- Spraying without protection;
- The reuse of pesticide containers, especially for preserving food and water intended for consumption;
- The storage of pesticides in dwellings;

The use of chemicals for other purposes (to kill head lice, wood weevils, rats, caterpillars, exotic species found in the vicinity of concessions).

The various effects of cotton production are contributing to the disappearance of natural resources. In addition to the negative impacts of cash crop production, farmers and COTONTCHAD-SN are facing difficulties that are leading to a decline in cotton production.

III - DIFFICULTIES LINKED TO PRODUCTION AND MARKETING

COTTON MARKETING IN CHAD

Cotton production, once the backbone of the Chadian economy in terms of cash income, is now undergoing a serious crisis, causing production to fall. This decline is due to difficulties encountered by farmers and the CO- TONTCHAD.

A - Difficulties linked to cotton production and marketing at farm level

In Chad, cotton growers face a number of difficulties during the growing season, preventing them from achieving higher yields.

1) Natural problems and difficult access to agricultural inputs and equipment

Problems related to nature are caused by the vagaries of the weather, which disrupt the agricultural calendar, with either delayed rainfall that does not favour early sowing, or abundant rainfall that floods sown areas. Soil degradation due to erosion and poor cultivation practices partly explain the major difficulties faced by farmers[373]. Indeed, the infertility of the soil as a result of these natural phenomena has led to a climate of mistrust among farmers, who are reluctant to express their needs when ordering inputs in order to avoid

---

[370] F. Nuttens, 2001, La production de coton graine dans la zone soudanienne, N'Djamena, Ministère de l'Agriculture, ONDR/DSN, p. 28.

[372] Interview with Allarayem, Ngandou, 22 June 2017

F. Nuttens, 2001, p. 34.

COTONTCHAD's debts. In addition, there has been a delay in the introduction of inputs at village association level. In 2006, less than 60% of village associations had received their orders for NPKSB fertiliser[374] and urea by 1er June, while 20% received their orders by the 15 June deadline for spreading. In addition, the quantities of NPKSB and urea actually applied remain 30% and 40% below the quantities ordered[375] .

Delayed deliveries of inputs are a handicap to agricultural development, which is so worrying for cotton growers. The partial fulfilment of input orders contributes to non-compliance with fertiliser doses, with repercussions on yields and incomes. The high cost of inputs and agricultural equipment means that growers are unable to expand their farmland in the Mandoul Oriental region[376] .

This shows the lack of a policy of subsidising the price of agricultural equipment by the State for the benefit of farmers, who have played an important role in ensuring food self-sufficiency throughout the country.

2) Lack of support for cotton growers

Cultivation techniques have changed little since cotton became widespread in the 1980s. While ploughing at the start of the cycle has become widespread, mechanical weeding remains marginal, and mechanical sowing is totally absent. However, the only supplier of mineral fertiliser is COTONTCHAD through the input credit scheme. The production and application of organic manure is still not widespread. Average cotton yields in Chad in 2005 and 2006 were low compared with those in other franc zone countries, where they exceeded one tonne per hectare. The low use of inputs remains one of the main reasons for the low yields in co-ton[377] . The limited technical and economic performance of Chad's cotton-growing zone does not encourage producers to repay input loans, and is a risk factor feared by members of village associations. In fact, the debts of some village associations have reached such high levels that repayments can be as high as the entire income from sales.

We also mentioned the poor management of land by producers on the grazing lines of transhumant herders in certain villages. The destruction of cotton fields is the cause of agro-pastoral conflicts in the region, with harmful effects on the population and a drop in production. After the hard work and difficulties encountered by the farmers, they experienced delays in the self-managed market and payment.

3) Delays in collection and payment to producers

Delays in the collection and evacuation of seedcotton from growers are the major difficulties noted, causing a drop in cotton production in the zone. At the end of August 2006, in the middle of the agricultural season :

12,000 tonnes of seedcotton produced in 2005 were still waiting to be evacuated, equivalent to more than 10% of overall production. At the same time, 20% of the village associations surveyed still had an average of 38 tonnes of seed cotton awaiting collection, equivalent to more than 7% of the production of the sample surveyed[378] .

The delay in "collecting seedcotton from village associations (VAs) has led some producers to take their cotton stock to neighbouring villages, where self-managed markets are open"[379] . This is characterised by a lack of transparency or non-compliance with the rules governing the

[374] NPKSB; type of complex fertiliser used in the cotton-growing zone of Chad.

[375] *Ibid.*

[376] Interview with Djimrangaye, Koko, 24 June 2017

[377] D. Reoungal, 2008, *Diagnostic agraire en zone cotonnière du Tchad, cas du village de Nguette*, Paris, L'Harmattan, p. 87.

[378] Interview with Nanhotomadje, Ndangtori, 20 June 2017

[379] Interview with Djim-hotongue, Koli, 20 June 2017

opening of self-managed markets.

However, producers complain that the market opening agreement is often extended for months after the harvest. This delay is unfounded and leads to a drop in the quality of the cotton because the producers do not have suitable premises for storing the cotton. According to the farmers, the estimated short distance between villages very close to the ginning plant has nothing to do with this, as it is common for these villages to open their markets months after the cotton has been harvested[380] .

However, VAs who are far from the factory sell their cotton just after the harvest or even during the harvest. Added to this is the corruption that keeps cropping up around the time of the cotton markets. In the villages where the administrative authorities produce cotton, the opening of the market is not long in coming, and neither is payment. So, for the cotton market to open up

In some village associations, farmers are required to make a contribution[381] of a sum of money paid to the cantonal delegate for the negotiation of the agreement to open up the self-managed market.

Another difficulty is linked to the devaluation of cotton quality, also known as the unfair "downgrading" of cotton. In the village associations in the Koumra cotton zone, many producers have testified to the reasons put forward by CO- TONTCHAD-SN for placing cotton in an inferior category: "the poor quality of inputs, inadequate storage conditions for seed cotton, wild fires, failure to respect the cotton harvesting schedule and damp cotton"[382] . They also know that mixing higher quality cotton with lower quality cotton in the same crate during transport to the ginning plant can cause the entire contents of the crate to depreciate.

However, for some producers, even if the cotton is "white as snow, it will always be downgraded"[383] . Once the cotton has been collected at the buying centre, it is transported to the factory, where it is assessed according to its quality. What growers still do not accept, however, is the unofficial manoeuvring that takes place during the grading procedure, resulting in their cotton being downgraded. An honest assessment of the quality of the cotton is generally the result of a private payment between the factory's cotton inspectors and the conveyors. This is why cotton growers often shout: "Just knock on the right door and the village cotton will not be downgraded"[384] . This tells us about certain villages where production is very high, but which are declared "not solvent" because of cotton downgrading practices. This means that farmers are only suffering because the reduction in income increases their vulnerability. They are faced with famine and the exorbitant interest rates offered by pawnbrokers. As a result, some farmers and village associations are forced to give up cotton production.

We note that late payment is a recurring problem for cotton growers in Chad. These delays do not encourage farmers to produce cotton. In the canton of Bessada in the east of the Mandoul Oriental region, "cotton growers have not yet received their cotton money for the 2015-2016 and 2016-2017 seasons"[385] . This has had a number of negative consequences for farmers, leading to discontent and the non-practice of cotton production in this part of the Mandoul Oriental region. Cotton growers say that before the self-managed market, cotton was paid for at the time of purchase in what is known as the ordinary market, so there were no late payments. Today, however, "after the market, COTONTCHAD-SN forgets that the cotton has

[380] Interview with Madjeram, Koli, 20 June 2017
[381] Interview with Madjeram, Koli, 20 June 2017
[382] Interview with Deouyo, Koumra, 14 June 2017
[383] Interview with Bolnan, Begue, 16 June 2017
[384] Interview with Bolnan, Begue, 16 June 2017
[385] Interview with Djimadjim, Bessada, 07 July 2017

been produced[386] , and producers receive their money when COTONTCHAD-SN has completed its seed cotton collection, evaluation and grading activities". This clearly shows that payment deadlines are indeterminate. This considerably increases food insecurity, as the majority of producers rely on cotton money to buy cereals and also to prepare for the next cropping season.

Many producers have preferred to devote all their resources to cotton production, obviously in order to pay for human labour and animal traction[387] . For the time being, it is the large producers with the financial means and material equipment who are producing cotton in large quantities in order to achieve their long-term projects.

However, "the majority of farmers have abandoned cotton production in favour of food crops to cover their daily needs. They feel that there is no comparison between the commercial crop known as cotton and food crops, which always remain in the family"[388] . Farmers rely on food crops, trade, livestock and fishing as their sources of income. In addition to the various difficulties, there is the lack of support for cotton growers in Chad, which is still a cause for concern, particularly in Mandoul. These are not only at the level of producers, but also at the level of COTONTCHAD-SN.

C - Difficulties encountered by COTONTCHAD-SN

The Chad cotton company, like others in Africa, was considered to be the lungs of the Chadian economy in the 1980s, but has encountered enormous difficulties in recent years. These are financial, structural and logistical in nature.

1) Financial difficulties encountered by COTONTCHAD-SN

In recent years, COTONTCHAD-SN has experienced cash flow difficulties due to economic factors. These have arisen as a result of fluctuating world prices, operating deficits and the debts of village associations and the state. However, the devaluation of the FCFA in 1994 gave a boost and the profits of cotton companies[389] , the need for reform became less urgent. COTONTCHAD experienced a very low level of cotton production just after the period of world price crises.

However, cash flow pressures have forced COTONTCHAD to limit its activities, including the shortage of inputs and the inability to purchase seed cotton in full. With a view to reversing this bad trend, COTONT- CHAD has set the purchase price of seed cotton to producers at 190 FCFA/kg for the 2005/2006 season, i.e. an increase of 30 FCFA over the period in line with the current situation on the world market[390] . To fulfil this promise, COTONTCHAD- SN had to request a subsidy of 10 billion FCFA from the State on the basis of an estimate of 25,000 tonnes of seed cotton, which covered only part of the actual amount[391] . In addition, the poor implementation of regulations relating to value added tax (VAT) in the cotton sector is one of the financial difficulties. This shows the State's tendency to circumvent or manipulate the law in its favour, penalising companies in the formal sector[392] . The value-added tax system in the cotton sector violates the principle of tax neutrality, because contrary to customary practice, the State does not refund the VAT levied on

386 Interview with Nguemadjibaye, Bessada, 07 July 2017
387 Interview with Kosramadje, Mousminda, 26 May 2017
388 Interview with Ra-adoumadje, Mousminda, 26 May 2017
389 F. Nuttens, 2001, p. 103.
390 E. Mbainaissem, 2013, p. 68.
391 Haroun Wawe, 2015, p. 36.
392 *Ibid.*

agricultural inputs. The high rate of this tax means that CO- TONTCHAD-SN is unable to cover its needs as it goes through financial crises[393] . In addition to these difficulties, COTONTCHAD-SN faces structural and logistical problems.

2) Structural and logistical difficulties

With regard to structural difficulties, the fall in cotton production in Chad is often attributed in part to the deterioration in structural factors. The decline is also due to the ineffectiveness of extension services and technical research in selecting seed generations. The cotton tracks have deteriorated and require serious work to restore them, and COTONT- CHAD-SN does not have sufficient financial resources for this work[394] . In fact, where the tracks are worked on by the State, rain barriers are installed as soon as the first rains arrive, and the barrier guards refuse to allow poly tippers to circulate. Meanwhile, COTONTCHAD-SN is in the middle of a marketing campaign to collect seed cotton from village associations.

The sale of inputs to producers is also a major problem at the Koumra factory. The problem remains unresolved because of the direct involvement of certain traditional chiefs, including the cantonal delegates, who have not renewed the office since their elections[395] . It should be pointed out that COTONTCHA-SN is experiencing difficulties due to the obsolescence of its plant and equipment, inefficient maintenance leading to recurrent breakdowns of trucks and machinery, and a lack of spare parts. Irregular seed deliveries, insufficient lorries for collecting seed cotton and evacuating cotton lint abroad, and high transport costs are forcing the Chadian cotton company to produce cotton in sufficient quantities compared with the average in some Central and West African countries[396] .

In addition, the consumption of electricity and fuel for maintenance is very high, and there are insufficient storage facilities for bales of cotton fibre, which are stored in the open air within the company.

As a result, the working conditions of the plant's permanent, seasonal and temporary staff are appalling. For example, the offices of some departments are too small for archives. The supervisor's office is right next to the factory's cyclone, where the gin sends out dust covering the documents, and all he did was vacuum every day. Staff lacked safety shoes and mackintoshes, putting their health at risk. The company cannot function as long as the human and material manpower is limited supervision[397] .

In short, cotton growing has had both positive and negative impacts on farmers since its introduction. It was seen as a source of revenue for the socio-economic development of Chad, and more particularly the Mandoul Oriental region. Indeed, access to inputs and agricultural equipment for cotton production has benefited the development of food crops through the after-effects of chemical fertilisers. This crop made possible an environmental development policy, livestock farming and infrastructure for social well-being. This crop was the real cause of the weakening of the Chadian economy, to the point where producers lost interest in producing cotton. In fact, the drop in the purchase price of seed cotton, the high cost of inputs and agricultural equipment, and the delay in collection and payment are fundamental reasons for the abandonment of seed cotton production, including the drop in the area sown and the tonne of production. At the same time, COTONTCHAD is experiencing difficulties in

[393] *Ibid*, p. 42.
[394] End of season report 2015/2016, p. 5.
[395] Interview with Deouyo, koumra, 09 June 2017
[396] Interview with Deouyo, koumra, 09 June 2017
[397] Interview with Deouyo, koumra, 09 June 2017

producing cotton fibre in sufficient quantities, and financial difficulties following the fall in world prices to pay producers and ensure logistics and personnel.

GENERAL CONCLUSION

This study looks at cotton production and marketing in Chad, focusing on its development from the colonial period to 2016. The colonial administration's cotton policy was set out in the 1928 convention commerciale du coton (cotton trade agreement) in the north of the AEF, which brought together four private companies, each of which was responsible for its own area of exploitation. The Chad cotton zone was entrusted to COTONFRAN, a company with majority Belgian and Dutch capital.

To this end, it had to ensure the exploitation of the area by distributing seeds to the farmers and ginning the harvested cotton, which was then exported to Le Havre to develop the French cotton industries. This explains why the colonial administration was obliged to force the population to take an interest in cotton production and had to set up agronomic technical research centres to train the supervisory staff who would follow the farmers through the cotton production system. This training phase was not easy because Chad's lack of appropriate basic infrastructure led the colonial administration to undertake the development of cotton production without the necessary application of conditions and cultivation methods. Given the absence of technical education in this part of the AEF, the administration thought it best to adopt the system of collective cotton growing followed by the traditional authorities. From then on, the idea was to use the rope to measure the agricultural areas allocated to a group of individuals aged between 15 and 50 for the compulsory cultivation of cotton. This behaviour is proof enough that cotton growing complements other colonial requirements. For this reason, the population of the cotton-growing areas, particularly in the Mandoul Oriental region, tried to resist this brutal threat by spontaneous uprisings. These led to a positive outcome, because cotton growers were rewarded and benefited from technical support for the development of cotton growing, which could be applied to food crops.

The colonial administration set up animal-drawn cultivation and mechanisation with a view to the sector, which could undoubtedly lead to a revival in the intensification of cotton growing. The success of this relaunch depended essentially on the synergy of actions between the main players, each of whom played his or her role to the full. In southern Chad in general, and in the Mandoul Oriental region in particular, farmers have access to inputs and agricultural equipment provided by COTONTCHAD and ONDR, which are subsidised by the State for rural development. However, before the oil era, cotton was still a major source of Chad's economy, but it has undergone a major crisis in recent years. This has led to a fall in prices for cotton growers, and hence a decline in cotton production. The ONDR's withdrawal from supervision meant that COTONTCHAD had to join forces with AGRIS to train cotton agents, who were now responsible for monitoring farmers.

Farmers have seen their interests threatened by falling purchase prices, high input costs and delays in collection and payment. This is why some farmers have given up cotton production. Others continued to grow cotton in the hope of gaining access to inputs not only for cash crops but also for food crops.

In a rural context marked by poverty and under-equipment, these changes have subjected the populatio ns to multiple constraints and challenges that weaken their production systems and lead them to over-exploit natural resources for various purposes. However, the main concern of cotton production remains the management of agrarian space to improve the conditions for agricultural and pastoral development. In many cases, these activities compete for the use of natural resources, leading farmers to seek solutions when the future of one or other system is threatened. In reality, farmers are obliged to manage their land properly to ensure that it is used in a harmonious way that favours agriculture and livestock farming, but the problem boils down to the poor use of these resources by the players involved. This partly reflects the problems of land tenure in the region with the dynamics of farming and livestock rearing, while the search for new land leads farmers to occupy livestock corridors. In this respect, it is difficult to assess the importance of one of these activities in place of the other, since producers are socially united in the system, whereas pastoralists are dispersed in the transhumance system. We have understood that livestock farming is at the heart of the dynamics of rural land management, a development encouraged as a replacement for the cotton crisis. Cotton cultivation has played an important role in the country's development, but producers are currently experiencing a number of problems in the areas of production and marketing. COTONTCHAD-SN has experienced financial, logistical and structural difficulties, resulting in high internal and external transport costs. These problems are largely due to the fall in world fibre prices, which has weakened the Chadian cotton company.

SOURCES AND BIBLIOGRAPHICAL REFERENCES PHIQUES

I-Sources

1) Archive Sources

- Koumra commune report, 2014, Communal Development Plan.
- Report from the sub-prefecture of Koumra, 2009, Census of the population of Mandoul Oriental.
- Report by Chairman TOMBALBAYE, 27 August 1974.
- Ministry of Overseas Finance, A.E.F, Chad, B.D.I.C.
- FAO, 2004, Programme National d'Investissement à Moyen Terme (PNIMT) au Tchad.
- Secretary of State for External Affairs, 1968, Economy and Development Plan, B.D.I.C.
- Report on the action and development plan in Chad.
- Industrial production activity report, 1957-1965, Logone brewery factory and CYCLOTCHAD company.
- COTONTCHAD, Guide pratique des activités commerciales de coton graine.
- NPKSB: Type of complex fertiliser used in the cotton-growing area of Chad.
- World Bank Washigton, DC, 2008, Final report on the organisation and outlook for African commodity chains: lessons from the reforms.
- Animal vaccination report 2015, Ministry of Livestock in the Mandoul Oriental region.
- COTONTCHAD-SN and Ministry of Agriculture, 2016, Rapport d'activités agricoles.
- Direction Huilerie Savonnerie, 2012, Production and marketing activity report.
- Ctrc, 2006, Rapport de mise en place des Comités de Coordinations Locaux.
- Ctrc, 2006, Réforme de la filière coton au Tchad. Progress report.
- ITRAD, 2007, Rapport d'activités de la production semencière du coton au Tchad.
- Rapport (Tchad), 1971, Protocole D'accord entre COTONTCHAD, ONDR, IRCT, DGRHA, DRHFRP et FIR.
- Société Cotonnière du Tchad, 2011, Presentation of the restructuring plan COTONTCHAD-SN (Société Nouvelle).
- Document de stratégie de reformes du secteur coton tchadien, 1999: adopté par le Haut Comité International. Republic of Chad, HCI Technical Committee.
- Practical guides to self-managed markets.
- Commercial activity report, Direction Huilerie Savonnerie Moundou (Chad).
- COTONTCHAD, 2009, Rapport des campagnes agricoles et commerciales.

3) Oral sources

N°	Full names	Age	Gender	Profession	Ethnic group	Date of interview	Maintenance site
1	Adamou	46	M	Sample weigher for cotton fibre	Gourane	14-06-2017	Koumra
2	Adoumngué	49	M	Teacher	Sara	21-06-2017	Koumra
3	Alladoumngue	35	M	Fire picket CT-SN	Sara	11-06-2017	Koumra
4	Allangombaye	35	M	Student	Sara	15-07-2017	Koumra
5	Allarayem	46	M	Retailer	Sara	22-06-2017	Ngandou
6	Allarayem	58	M	Driver CT-SN silo	Nar	13-06-2017	Koumra
7	Beramgoto	56	M	Local	Goulaye	08-06-2017	Koumra

				coordinating agent CT-SN			
8	Bolnan Elias	47	M	Cultivator	Sara	23-06-2017	Kemkian
9	Deouyo	53	M	Supervisor CT-SN	Ngambaye	07-06-2017	Koumra
10	Didjenbaye	52	M	CT-SN Input Warehouseman	Goulaye	09-06-2017	Koumra
11	Djadimadje	42	M	CT- Worker SN	Sara	07-06-2017	Koumra
12	Djainadaye	48	M	Cultivator	Sara	09-06-2017	Koumra
13	Djasngombaye	53	M	Cultivator	Sara	23-06-2017	Kemkian
14	Djimadjim	24	M	Student	Sara	07-07-2017	Koumra
15	Djimhotongué	38	M	Cultivator	Sara	12-06-2017	Koumra
16	Djimhoudjeta	23	M	farmer	Sara	16-06-2017	Kotkouli
17	Djimrangaye	47	M	Cultivator	Sara	24-06-2017	Koko
18	Djimtolabaye	51	M	Shift manager CT-SN	Sara	14-06-2017	Koumra
19	Guidimbaye	58	M	Mayor	Goulaye	19-06-2017	Koumra
20	Hoiadi Sidonie	43	F	Farmer	Sara	16-06-2017	Kotkouli
21	Kosramadje	46	M	Cultivator	Sara	26-05-2017	Mousminda
22	Madjadoum	38	M	Doctor	Gor	10-06-2017	Koumra
23	Madjeyenane	39	M	Fire picket CT-SN	Sara	11-06-2017	Koumra
24	Madjissembaye	46	M	Cultivator	Sara	24-06-2017	Koko
25	Maikoe	65	M	Cultivator	Sara	22-06-2017	Ngandou
26	Marmai	43	M	A P E F C	Gor	14-06-2017	Koumra
27	Mbaitoloum	34	M	Student	Ngabaye	12-07-2017	Koumra
28	Mbang-adoumbé	57	M	Chairman AV.	Sara	21-06-2017	Nderguigui
29	Mognara	57	M	farmer	Sara	24-06-2017	Ngomana
30	Mounmotoi	59	F	Housekeeper	Sara	18-06-2017	Bessada
31	Nanhotomadje	58	M	Cantonal delegate	Sara	20-06-2017	Ndangtori
32	Nantoyim-kingar	54	M	Village chief	Sara	16-06-2017	Kotkouli
33	Nassaryem	56	M	Cultivator	Sara	05-06-2017	Koumra
34	Neldissengar	54	M	Mayor Ad.	Goulaye	19-06-2017	Koumra
35	Ngar-idjimte	62	M	farmer	Sara	18-06-2017	Bessada
36	Ngaryedji	46	M	Ball ringers	Goulaye	14-06-2017	Koumra
37	Nguemadjibaye	53	M	Cultivator	Sara	07-07-1017	Bessada
38	Nguenanbaye	37	M	Cultivator	Sara	23-06-2017	Kemkian

39	Oumar	32	M	Bridge weigher CT- toggle SN	Zaguawa	14-06-2017	Koumra
40	Pafing Chiakré	54	M	RAC CT-SN	Moundang	12-06-2017	Koumra
41	Ra-adoumadje	58	M	Cultivator	Sara	26-05-2017	Mousminda
42	Ramadi	37	F	Housekeeper	Sara	21-06-2017	Kotkouli
43	Rassemadji	52	M	Cultivator	Sara	24-06-2017	Begue
44	Rimtoibaye	64	M	Head of sector ONDR	Goulaye	06-06-2017	Koumra
45	Roidji	22	F	Housekeeper	Sara	13-06-2017	Koumra
46	Sadjinan	37	M	Retailer	Sara	24-06-2017	Koko
47	Sillenengar	63	M	Cultivator	Sara	24-06-2017	Ngomana
48	Tchadingar	64	M	Cultivator	Sara	19-06-2017	Koumra
49	Telembaye	53	M	Plant manager CT-SN	Ngambaye	13-06-2017	Koumra
50	Vaitchiou	56	M	Head of cottonseed hulling workshop	Moundang	20-06-2017	Koumra
51	Y emingaye	64	M	Cultivator	Sara	22-06-2017	Ngandou
52	Y engar André	65	M	Teacher	Sara	20-06-2017	Koumra

II-Bibliography

1) Publications

> Anon, 1991, *Mémento de l'agronome,* Ministère de la Coopération et du Développement, 4e Edition, Paris, Collection "Techniques Rurales en Afrique".

> Berthelot, J., 2001, *La mise en boites du soutien interne ; Attention aux étiquettes mensongères*, Paris, Montpellier.

> Beyem, R., 2000, Tchad : *l'ambivalence culturelle et l'intégration nationale*, Paris, L'Harmattan.

> Bouquet, 1982, *Tchad, genèse d'un conflit,* Paris, L'Harmattan.

> Boussard, J., 2005, *Libéraliser l'agriculture mondiale, Théories, modèles et réalités,* Paris, Monpellier.

> Braud, M., 1990, *La filière coton en Afrique de l'Ouest et du Centre, un îlot de progrès dans un Océan de morosité*, Paris [FR]: CIRAD-IRCT.

> Buijtenhuijs, R., 1978, *Le Frolinat et les révoltes populaires du Tchad* (19651976), Paris, Mouton.

> Cabot, J. and DIZIAIS, R., 1955, Population du Moyen Logone (Cameroun et Tchad), "L'Homme d'Outre-Mer", Paris, O.R.S.T.O.M.

> Chevalier, P. S., 1949, *Le coton*, Paris, PUF.

> Diguimbaye, G. and Langue, R., 1969, *L'essor du Tchad*, Paris, PUF.

> Ela, J-M., 1994, Afrique : *l'irruption des pauvres. Société contre l'Ingérence, pouvoir et Argent*, Paris, L'Harmattan.

> Floret, F., 1993, *La jachère en Afrique tropicale*, Paris, Unesco, MAAB dossier.
> Georges, R., 1989, Le coton en Afrique de l'Ouest et du Centre. Situation et perspectives, Montpellier: CIRAD-MESRU.
> Hazard, E., 2005, *Enda prospectives dialogues politiques*, enda éditions, Dakar.
> Hechscher, E., 1972, *Echange international et croissant, Economica*, Paris, PUF.
> Jean-Louis, C., 2003, *Cultures vivrières et commerciales en Afrique Occidentale*; le ourd collection, question de géographie, Nantes (France), Éditions du temps.
> Kraft, J., 1999, *Le processus de concurrence,* Economica, Paris, Gallimard.
> Lemoinet, Tchad, 1960-1990, *Trente années d'indépendance,* Paris, Lettre du monde.
> Levrat, R., 1950, *le coton en Afrique Occidentale et centrale avant 1950*, L'Harmattan.
> Levrat, R., 1950, *le coton en Afrique Occidentale et Centrale avant 1950, un exemple de la politique coloniale de la France*, Etudes africaines, Paris, L'Harmattan.
> Madana Nomay, 2001, *Les politiques éducatives au Tchad,* Paris, L'Harmattan.
> Maurois, A., 1937, Dictionnaire Larousse, Paris, Gallimard.
> Montchretien, A., 1993, *Histoire des pensées économiques*, Paris, Sirey.
> Pairault, 1994, *Le retour au pays d'Iro, chronique d'un village du Tchad*, Paris, Karthala.
> Reoungal, D., 2008, *Diagnostic agraire en zone cotonnière du Tchad, cas du village de Nguette*, Paris, L'Harmattan.
> Robinson, J., 1993, *L'économie de la concurrence imparfaite*, Paris, Dunod.
> Sarraut, A., 1932, *La mise en valeur des colonies françaises*, Paris, Payot.

2) Theses

> Abakar Gouni Ousman, 2009-2010, "Le commerce extérieur du Tchad de 1960 à nos jours", PhD thesis, University of Strasbourg.
> Abakar Kassambara Abdoulaye, 2010, "La situation économique et sociale du Tchad de 1900 à 1960", doctoral thesis at the University of Strasbourg.
> Aboubakar Ali Kore, 2011, "La socialisation politique au Tchad. Analyse critique du contenu des livres Scolaires pour la période 1960-2005", PhD thesis, University of Franche-Comté.
> Kibassim Bagrim, 1975, "Agriculture commerciale, modernisation et développement rural en zone cotonnière, exemple du Moyen Chari", PhD thesis in Development Sociology, University of Paris V.
> Macra Tadin, 1983, "L'intervention de l'État dans le secteur cotonnier au Tchad", Doctoral thesis in Public Law, University of Toulouse.
> Magnant, J-P., 1983, "Terre et pouvoir chez les populations dites 'Sara' du Sud du Tchad: la famille, l'individu et l'Etat, leur terroir et leur territoire", PhD thesis in Political Science, University of Paris I.
> Magrin, G., 2001, "Le sud du Tchad en mutation : des champs du coton aux sirènes de l'or noir", doctoral thesis in geography, Université de Paris, Panthéon-Sorbonne.
> Nasura, H., 2002, "Les stratégies de développement et les politiques de sécurité alimentaire en Afrique subsaharienne. Le poids des incohérences", PhD thesis, Unité d'économie rurale, UCL.
> Reounodji, 2003, "Espaces, sociétés rurales et pratiques de gestion des ressources naturelles dans le Sud-Ouest du Tchad vers une intégration agriculture- élevage", PhD thesis, University of Paris I/Panthéon-Sorbonne.

3) Memoirs

> Armi, J., 2003, "L'économie cotonnière dans la région de Pala au Tchad (19252000)",

Master's thesis in History, University of Ngaoundéré.
> Djoubdje Alamine, 2013, "L'Impact de la culture du coton dans la Tandjile- Ouest (Tchad): 1930-2012", Master's thesis in History, University of Ngaoundéré.
> Mbainaissem, E., 2013, "Audit stratégique de la Cotontchad-SN", dissertation, Institut de N'Djamena.
> Wawe Haroun, 2015, "Pratique de contrôle budgétaire, Cotontchad-SN", Professional Master's thesis, University of Ngaoundéré.

4) Articles and periodicals

> Arditi, C., 2004, "Des paysans plus professionnels que les développements? The example of cotton in Chad (1920-2002)". *Tiers monde,* Year 2004, Volume 31, Number 180.
> Bigot, Y., Raymond G., 1991, "Traction animale et motorisation en zone cotonnière d'Afrique de l'Ouest : Burkina Faso, Côte-d'Ivoire, Mali", CIRAD- DSA, CIRAD-IRCT, *Collection Documents Systèmes Agraires*, No. 14, CIRAD.
> Cabot, J., 1957, "La culture du coton au Tchad", *Annales de géographie, volume 66, number 358,* C.A.O.M.P.
> Charriere, G., 1984, "La culture attelée : un progrès dangereux", *ORSTOM notebook, Sciences Humaines series*, volume 20, number 3-4.
> Clanet, J. C., 1982, L'insertion des aires pastorales dans les zones sédentaires du Tchad central. *Cahiers d'Outre-Mer*, vol. 35, n° 139.
> FAO, 2005, "The State of Food and Agriculture. Agricultural trade and poverty": *Can trade serve poverty*? Rome, FAO.
> Gaid, M., 1956, "Au Tchad, les transformations subies par l'agriculture traditionnelle sous l'influence de la culture cotonnière", Comité de Coordination de la Recherche Agronomique et de la Production Agricole. Gvt. Général de l'A.E.F.
> Gerald, E., 2006, "Le marché mondial du coton : Évolution et perspectives", *Cahiers Agricultures*, issue 1.
> Haessler, C., et al. 2002, "Développement du cheptel au Sud Tchad : quelles politiques pour l'élevage des savanes ?", In Jamin J.Y., Seiny Boukar, 2002, Les *savanes africaines : des espaces en mutation, des acteurs face à de nouveaux défis.* Conference proceedings, Garoua, Cameroon, N'Djamena, Chad.
> Lenfant, 1909, " le coton pousse à vue d'œil à proximité des cases par l'apathie de l'indigène, reste inutile et se perd ", *La découverte des Grandes Sources du Centre d'Afrique*, vol.37, N°54.
> Lhuilier, J., 1900-1950, Tchad, le " coton " *Tropique,* n° 328.
> Ngarlamulum, T. G., 2002, "l'organisation du travail agricole dans la ceinture verte de la ville de Kananga", *Annales de l'ISP-Kananga*, vol. XI.
> Pelissier, P., 1980, "L'arbre dans les paysages agraires de l'Afrique noire". *In l'arbre en Afrique tropicale, la fonction et le signe*, Cahiers ORSTOM, Série sciences humaines, vol. XVII.
> Peltre-Wurtz, J., 1984, "La charrue, le travail, l'arbre", *cahiers, Sciences Humaines series*, volume 20, number 3-4.
> Piguet, F., 2002, "Le concept de sécurité alimentaire", in N. S. Tircier and B. Sottas (eds.*), la sécurité alimentaire en question. Dilemmes, constats et controverses*, Paris, Karthala.
> Rameau, G., 1953, "L'élevage bovin au Tchad", *Revue internationale desproduits coloniaux et du matériel,* number 282, C.A.O.M.

> Roose, R., 1985, "Impact du défrichessement sur la dégradation des sols tropicaux", *tropical agricultural machinery,* issue 87.

> Seignobos, C., 1981, "L'arbre et la cité dans la zone soudano-sahélienne, exemple du Tchad et Cameroun", *Revue de géographie du Cameroun*, numé- ro1.

> Stevelinck, W., 1953, "Le développement du coton dans la zone Mayo Kebbi, Logone et Moyen Chari", *Marches coloniaux du monde,* number 399, C.A.O.M.

> Stiglitz Joseph, E., and Charlton, A., 2005, *"For a fairer trade",* O.U. Press, Oxford.

> Ulrich Stuizinger, Chad, 1983, "mise en valeur", *cotton and development, Tiers-Monde, Year 1983,* Volume 24, Number 95.

5) Electronic sources

> A. Maurois, 1937, Dictionnaire de l'Académie française, 8ème edition, Mercure de France, Paris, https://fr.wikionary.org/wiki/commerce, accessed 30 March 2017.

> A. Smith, 1776, The Wealth of the Nation, http://www.leconomiste.eu/descryptage-economie/211-idée-clef-de-la-richesse- des-nations-d-adam-smith,html, accessed 08 March 2017.

> B. Bathelot, 2015, "The illustrated encyclopaedia of marketing", https://www.google.com/search?q=marketing&ie=utf-8a&oe=utf-8, accessed, 26 June 2017.

> C. Araujo-Bonjan and J.F. Brun, 2000, Les politiques de stabilisation des prix du coton en Afrique de la zone franc sont-elles condamnées? Revised version of a paper presented at the conference "Dynamique des prix et des marches de matièrespremières : analyseetprévision ", http://www.agriculture.gouv.fr/spip/IMG/pdf/coton_-marche.pdf, accessed, 14 July 2017.

> F.Perroux, 1964, *L'économie duXXme siècle*, Paris, PUF, http://conte.U- bordeauX4.fr, consulted on 13 March 2017.

> Fille://D://coton-wikipedia.htm, consulted on 30 March 2017.

> http://www.dailymotion,com/video/Xcu89hproduction de-spiruline-au-lac-tchao lifestyle, accessed on 05 March 2017.

> http://www.dailymotion.com/video/Xcu89hproduction de - spiruline-au-Lac- tchao lifestyle, consulted on 09 May 2017.

> https://fr.wikipedia.org/wiki. History of cotton growing, accessed, 30 June 2017.

> https://www.aquaportail.com/definition_5841_developpement rural.htm, consulted on 30 June 2017.

> J.C Chebath, B. G. Simard, 1971, *Le vendeur méconnu dans connu*, www.cnrt/en/definition/marketing, consulted on 26 June 2017.

> J.J. Courtant et al, 1991, Le coton en Afrique de l'Ouest et du Centre, Ministère de la Coopération et du Développement, https://www.memoireonline.com/09/09/2706/m_Effet_de-differentes-pratique- de-t, consulted on 07 September 2017.

> LesitedesEditionsLarousse , www.larousse.fr./encyclopedie/divers/commerce/35477, accessed on 30 March 2017.

> M. Fok, 2006, "Crises cotonnières en Afrique et problématique du soutien", BASE [Online], volume10, number4, Biothenol. Agron. Soc. Environ, URL:http://pops.ulg.ac.be/1780-4507/index.php ?id=562, accessed on 07 September 2017.

> M. Sorto, 2006, L'amélioration de la qualité de Dihé, la spiruline récoltée au Tchad, Spirulinagadez.free.fr/0503pm.htm, consulted on 05 March 2017.

> P. Artus, 1998, "Are French companies going to start up again? s'endetterter?", *Revue d'économie financière n°46*, Endettement/Surendettement, https://www.jstor.org/stable/42903599? Seq=1#page_scan_tab_contents, accessed 27 March 2017

> Site Transports- Ministère de l'écologie, du développement et de l'aménagement durables: http://www.transports.equipement.gouv.fr, consulted on 26 June 2017.

REPUBLIQUE DU CAMEROUN
Paix - Travail – Patrie

UNIVERSITE DE NGAOUNDERE
B.P. : 454 Ngaoundéré
E-mail : rectorat_ngaoundere@yahoo.fr

FACULTE DES ARTS, LETTRES ET SCIENCES HUMAINES

REPUBLIC OF CAMEROON
Peace - Work - Fatherland

UNIVERSITY OF NGAOUNDERE
P.O. Box: 454 Ngaoundéré
E-mail : rectorat_ngaoundere@yahoo.fr

FACULTY OF ARTS, LETTERS AND SOCIAL SCIENCES

Le Vice-Doyen chargé de la Recherche et de la Coopération
The Vice-Dean in charge of the Research and Cooperation

N° 159/17 /UN/DFALSH/CDH

Ngaoundéré, le 29 MAY 2017

ATTESTATION DE RECHERCHE

Le Vice-Doyen chargé de la Recherche et de la Coopération de la Faculté des Arts, Lettres et Sciences Humaines, de l'Université de Ngaoundéré, atteste que l'étudiant **BEYENAN NGARASNDI**, né vers 1990 à KINDIRI (Tchad), est inscrit en MASTER Recherche, au titre de l'année académique 2016-2017 dans la filière Histoire Economique et Sociale, suite à la décision **N°2016/331/UN/R/VR-EPDTIC/DAAC/DFALSH/VDSS**, sous le matricule **12B178LF.** Il effectue, dans le cadre de son Master, un travail de recherche sur le thème : **« La production et la commercialisation de coton au Tchad : cas de Koumra de 1949 à 2016 ».**

Nous le recommandons auprès des institutions, organismes et personnes ressources susceptibles de lui fournir des informations nécessaires à la réalisation de son étude.

En foi de quoi, la présente attestation lui est délivrée pour servir et valoir ce que de droit.

Le Vice-Doyen

Dr. Oaba Abdoul-Bagui
Chargé de Cours

COTONTCHAD SN

BORDEREAU D'EXPEDITION

N° [illegible]

Magasin Expéditeur : Code : DESTINATAIRE Code :

N° Véhicule : CCI : Code AV :

R	Nomenclature	Désignation	U	Qté	Prix Unitaire	Valorisation	RESERVES
1							
2							
3							
4							
5							
6							
7							
8							
9							
10							

Magasinier Expéditeur	Le Chauffeur pour Prise en Charge après Contrôle	Réceptionné à	Réceptionné à
Nom :	Nom :	Par :	Par :
Prénom :	Prénom :	Fonction :	Fonction :
Date :	Date :	Date :	Date :
Signature :	Signature :	Signature :	Signature :

COTONTCHAD SN

BORDEREAU DE RECEPTION N° 0042349 ☐ Externe ☐ Interne

N° Véhicule : ____________ Nom Chauffeur : ____________ Transporteur : ____________

N° Commande : ____________ Fournisseur : ____________ Bordereau de Livraison N° ____________ en date du ____________

N° Bulletin de Mouvement : ____________ en date du : ____________ Magasin Expéditeur : ____________ Code |___|___|

Magasin Réceptionnaire : ____________ Code ____________

R	Nomenclature	Désignation	U (1)	Qté	Observations	Valorisation (Comptabilité)
1						
2						
3						
4						
5						
6						
7						
8						
9						
10						

(1) Doit être identique à celle des fiches de gestion des stocks magasin

Le Responsable d'Activité Pour Visa	Le Magasinier pour Prise en Charge Après Contrôle	RESERVES
Nom (2) ____________ Prénom (2) ____________ Date (2) ____________ Signature (2) ____________	Nom (2) ____________ Prénom (2) ____________ Date (2) ____________ Signature (2) ____________	
		Exemplaires : 1-DFC, 2-Direction/Service Concerné, 3-Responsable du Magasin, 4-Transitaire, 5-Souche

(2) Obligatoire

Printed by Books on Demand GmbH, Norderstedt / Germany